Ducks and How To Make Them Pay

by William Kinard Cook

with an introduction by Jackson Chambers

This work contains material that was originally published in 1899.

This publication is within the Public Domain.

This edition is reprinted for educational purposes and in accordance with all applicable Federal Laws.

Introduction Copyright 2018 by Jackson Chambers

The World's Largest Selection of Vintage Poultry Books

www.VintagePoultry.com

Self Reliance Books

Get more historic titles on animal and stock breeding, gardening and old fashioned skills by visiting us at:

http://selfreliancebooks.blogspot.com/

Introduction

I am pleased to present yet another title on Poultry.

The work is in the Public Domain and is re-printed here in accordance with Federal Laws.

As with all reprinted books of this age that are intended to perfectly reproduce the original edition, considerable pains and effort had to be undertaken to correct fading and sometimes outright damage to existing proofs of this title. At times, this task is quite monumental, requiring an almost total "rebuilding" of some pages from digital proofs of multiple copies. Despite this, imperfections still sometimes exist in the final proof and may detract from the visual appearance of the text.

I hope you enjoy reading this book as much as I enjoyed making it available to readers again.

Jackson Chambers

From a Photo. by Mr. E. Davey Lavender, Bromley, Kent.

WILLIAM COOK.

CONTENTS.

	PAGE
PREFACE	1—3
INTRODUCTION	5—7
GENERAL INFORMATION	9—15

Ducks pay well—Stock ducks; water or no water—Selection for Show Birds—Freaks of nature—Ducks in the garden and fields—Ducks as vermin destroyers.

STOCK DUCKS	17—30

Management—Stock ducks and ducklings—Laying ducks—Meat diet for egg producing—Warm water for cold weather—The duck house: construction and litter, warmth and ventilation

SITTING AND HATCHING	31—38

Setting—Care of eggs during incubation—Duck eggs and incubators—The nest: formation, and material to be used—Egg testing to ensure fertile eggs being sat upon.

FEEDING AND MANAGEMENT OF YOUNG DUCKS	39—52

Rearing Ducklings—Feeding young stock—Stock and exhibition ducks—Ducks for home consumption—Ducks for market—Aylesbury duck management—Duck houses and barns—Fattening—Warmth for growth.

CONTENTS.

	PAGE
HOUSING YOUNG DUCKLINGS	53—56

Houses for ducklings—Penning in open air—Ducks in orchards—Windows and ventilation for duck houses.

PURE-BRED DUCKS—AYLESBURY STOCK DUCKS ... 57—79

Aylesbury ducks—Points of the Aylesbury—Duck eggs *versus* chickens' eggs—Stock Aylesbury ducks and trickery—Good breeders for good ducks—Pens of Aylesbury Ducks: how to select and start—Rouens: their characteristics and crosses—Pekins: their points and peculiarities—Cayugas: their antecedents and merits—Muscovies: their habits and heaviness—Black East Indian, "little and good," their points and perfections.

CROSS-BRED DUCKS... 81—92

Crossing for table a profitable thing—The old and the new crosses: comparative advantages—Indian Runner crosses for laying—Good feeding for good egg results—Pekin-Rouen—Muscovy-Aylesbury—Wild-Rouen.

THE HISTORY, SPREAD AND DEVELOPMENT OF INDIAN RUNNERS... 93—97

Indian Runners—Laying points of Indian Runners—Indian Runners for crossing.

WHITE INDIAN RUNNERS ... 99—102

New breeds introduced—Reception of new breeds by captious and badly-informed people—Points of White Indian Runners—Laying qualities and characteristics of Indian Runners generally—Beware of fraudulent imitations—Activity of Indian Runners.

FATTENING ... 103—106

Fattening an important part of the duck industry—Food that will fatten—How to treat ducks when fattening—Killing: how to select for—Shelter for ducks.

CONTENTS.

	PAGE
GRIT	107—112

Sharp Grit for Ducks: let it be sharp—Flint Grit *versus* Sand, Shell, and Shingle—Aylesbury Ducks: fallacies and facts—Dangers arising from insufficient sharp grit.

How to make Duck Ponds ... 113—120

Duck Ponds should be substantially made—Cement, Concrete, and Ornamental Duck Ponds—Cheap Substitutes for a properly constructed Pond—Ponds for Stock Ducks: Fertile Eggs or Unfertile Eggs.

Diseases ... 121—126

Ducks not subject to many ailments—Treatment for cases of consumption and oviduct displacement—Roup, staggers, and loin weakness in ducks: how to deal with.

Supplementary Chapter ... 127—128

Ducklings for the multitude—The plan impracticable—Present Position—Eggs and Ducklings: Counsel and Advice.

INDEX.

	PAGE
Aylesbury Duck Breeding	45
Aylesbury Breeders: plans for feeding and rearing	46
Artificial heat for ducklings	54
Aylesbury Ducks	57
,, ,, laying properties of	59
,, ,, how to procure good stock	60
Black East Indian: their uses and points	79
Cayugas: their history and probable extraction	70
Cross-bred ducks: laying qualities of	85
Duck-rearing industry in England, The	6
Duck-keeping for profit	9
Duck ponds	10, 113
,, materials for	114
,, drainage and cleaning out	117
,, ornamental	119
,, inexpensive forms of	119
Ducks as scavengers in the garden	14
Ducks in cold weather: management of	26
Ducks *versus* hens for hatching duck eggs	32
Ducklings	12
,, how to produce fine	39
,, how to treat immediately after hatching	40
,, how to teach to feed	41
,, food for	42
,, for stock	43
,, milk for	49
,, meat for	22
,, sunstroke	49
,, outdoor pens for	48
,, artificial heat for	54
,, rice for	49
,, marketing, management for	49, 51

INDEX.

	PAGE
Drinking water for fat ducklings	44
Duck barns	45
Duck house flooring	53
Diseases	121
,, treatment for	122
Egg passage displacement: treatment for	123
Feeding for egg production	13
Feeding: general	24
Flint grit	25, 107
,, *versus* shingle and sand	110
Fattening	103
Food for fattening purposes	104
Grain for ducks: preparation of	25
Home fed and foreign ducks: comparative advantages of	5
Hatching: how best to manage	34
,, how to liberate ducklings	36
Houses for ducks: shapes and sizes	28
,, ventilation	29
,, how to disinfect	30
Incubators	32
Incubation: directions for procuring successful	33
Incubators for large breeders	38
Indian Runners, history and development	93
,, laying qualities of	94
,, points of	96
,, for crossing	82, 96
,, white	99
,, points of white variety	100
Inflammation of lungs	126
Moss peat and straw litter	27, 48
Moulting	43
Muscovy-Aylesbury, incubation of	90
Muscovy drakes, description of	74
,, ducks: peculiarities and inferior laying qualities	74
,, ducks and drakes: weight and peculiar carriage	77
Open sheds for feeding and sheltering young ducks	55
Orchards for duck rearing	55
Pekins: their peculiarities and advantages	68
,, and rats	69

INDEX.

	PAGE
Pekin-Aylesbury cross: size and prices	89
Pekin-Rouen cross, points of	87
Roast ducks for the multitude	128, 5
Rouens: their points and great beauty	64
Roup	123
Rouen-Aylesbury cross: how to utilize and improve	81
Selection for breeding: what birds to avoid	12
Stock ducks, general treatment of	11
,, treatment of when laying	17, 20
,, moulting	18
,, catching	20
,, water for, how to arrange	19
,, meat for	22
Sitting and hatching	31
Staggers	125
Soft food, how to prepare	23
Ventilation	54
Windows in duck houses	56
Wild-Rouen cross, how to procure	91
Young ducklings, supply of	127

LIST OF ILLUSTRATIONS.

Aylesbury drake	56
,, duck	58
Black East Indian duck and drake	78
Cayuga duck and drake	70
Indian Runner drake	94
Muscovy duck and drake	74
Pekin duck and drake	68
Portrait of Mr. Cook	Frontispiece
Rouen duck and drake	64
Rouen Aylesbury cross	82
White Indian Runner drake	98
,, ,, duck	100

PREFACE.

THE original idea which led me to publish the first edition of this book was that there was a need for sound practical information upon a subject, which, if properly taken up, would be likely to lead to great results being attained in connection with an industry which would not only provide employment and remuneration for many, but also open up a way for families, both in town and country, to produce ducks for their own consumption, even where only the smallest accommodation existed. Ducks should form a great part of our bill of fare from April to October. I have endeavoured to show in a simple form how they can be kept both as stock ducks in small places and how the young ones can be reared easily, as ducks grow faster than chickens, and are much less trouble. I have not gone into the treatment of fancy ducks, but have given a description of all the most useful breeds, and how they may be kept under different circumstances; so that readers may judge at once which breed is best suited to their purpose. I have not only treated on pure ducks, but have shown how these can be crossed to the best advantage. The best methods of housing and feeding are described and the whole management in detail, so that those who have never kept ducks may see very clearly how they should set to work at once and rear ducks for their own table. I have also given directions for the management of exhibition stock ducks to guide the young amateur, as well as a few hints to the practical and experienced duck breeders. I may say the little book is written on practical experience of my own, as I have kept all breeds which I have mentioned and have proved their merits. I have also shown how all the refuse from the table can be turned into good and profitable food by keeping a few ducks. Good fat young ducklings have been and are to a very great extent luxuries for the rich, while the majority of people do not know the taste of a good home-fed young duck. There are thousands of foreign ducks imported from various parts of the world which find their way into London and other large markets in England. These are of a very inferior kind, with scarcely the flavour of a duck in them, and at the same time they are tough to eat. They give those who eat them a very bad impression of what young ducklings ought to be. They are also very small. Many of them do not weigh more than 3 lbs. each. There is no reason why we should not produce in England one hundred times more ducks than are reared at the present day, as quite as many ducks as would be necessary could be reared in this country to supply the markets and our own consumption as well. Duck-rearing should become one of our industries, and I have every reason to believe that it will in a few years to come.

The book has now been revised, and brought up to date, with much additional information, which it is hoped will help duck keepers to become more successful in the future, even than many who have been successful in the past.

W. COOK.

ORPINGTON HOUSE,
St. Mary Cray, Kent.

INTRODUCTION.

NOT so many years ago roast duck was a luxury only for the rich, unless it might have been a few foreign birds sent into our markets, which could usually be bought up at a very low price. Many of those who were purchasers of these foreign ducks, when they have had the task of carving and eating them, shook their heads and said if that was roast duck a little of that would go a very long way. The foreign birds were not only very poor, and not more than half fed, but there were many old ones put in amongst the young ones. Therefore those who were really in the habit of buying ducks, did not know the taste of home-bred and home-fed ducks, for there is no comparison between these and the foreigners. I have written many articles on how cottagers and those who have only a very limited accommodation could easily hatch and rear ducks to produce food for their families. I am pleased to say that hundreds have carried out my instructions, and have been able to produce their home-fed ducklings as food for the family, and not merely as a luxury. Young ducklings grow up so fast, and require so little attention, that it is a very easy matter to rear one's own. I believe that there are one hundred young ducklings reared at the present time, in various parts of the country, where there were only ten reared some years ago, and soon there will be one thousand reared where there are one hundred produced at the present time. Young ducklings should really form a great part of our daily food, and especially where there are large families. The food they thrive on is cheap, and all kinds of butcher's meat is dear; not only so, but it is far more satisfactory to be able to produce flesh meat for one's own table, than to buy elsewhere. It has been thought by thousands of English people that ducks were very difficult to rear, and that it was only those who lived in the country and had a large pond, river, or brook for them to swim in who were able to rear them at all, but I am pleased to say that I have proved this to be entirely wrong. Ducks can be reared in towns just as well as in the country, and in a limited space without water, except just enough for drinking purposes only, better than they can in a farmyard, or where they have a large pond or stream to go to. They grow faster, and fatten quicker if they are never allowed to go into the water at all. They do not require an elaborate house; a large box, made watertight, or a small shed is quite sufficient for them where one only wants to rear a few for the table. I have dealt very briefly in this little work on how ducks can be managed on a small and large scale, both for exhibition and market purposes; which are the best breeds to keep, as regards pure and cross breds, according to circumstances; which are the best and cheapest houses to keep them in; and how to distinguish the best ducks for stock, as well as giving rules for rearing and fattening the young stock. I have every reason to believe that a new and revised edition of this little work will be taken up very largely, and if it goes anything like "The Poultry Breeder and Feeder: or, How to Make Poultry Pay," which has the largest circulation of any poultry work in England, and from which many have derived great benefit, I shall feel that I have done something towards bringing about a cheap food for the people of England. It must be understood that a duck's egg is worth more than a chicken's egg, as it is much larger, and many say it is more nutritious. At any rate, it is a change from a hen's egg, and some delicate people can eat a duck's egg when they cannot touch the former. I hope those into whose hands this edition of the book may fall will make it widely known, for by so doing I feel sure they will be doing their neighbours and friends a good service. Whatever may have been omitted in any chapter of this book I shall be only too pleased to give any further information upon to any one. I always give advice on all poultry matters, to any one sending me a stamped and addressed envelope, and I am willing to give advice also on the management and feeding of ducks. I shall be pleased to answer all questions as far as I am able.

Duck rearing has become a great increasing industry, and it is to be hoped that the English people may realise more and more what a grand field for enterprise this question opens up, and what a large amount of money might be kept in the country if only our cottagers and farmers aroused themselves and applied their energies to its retention.

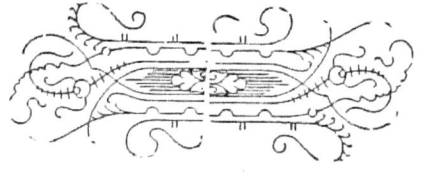

GENERAL INFORMATION.

Ducks pay well—Stock ducks; water or no water—Selection for Show Birds--Freaks of nature—Ducks in the garden and fields—Ducks as vermin destroyers.

DUCKS in many cases pay their way better than fowls, and especially when they are only reared for the table, as they grow so quickly and can be reared without water to swim in. Some of my readers may ask, "How about stock ducks? Will they breed without water?" Yes; ducks' eggs will be fertile if the stock birds are never allowed to go in the water at all; but when they are once allowed to go into the water, and then taken away from it, the eggs may not be quite so fertile for a time, as the water is the natural place for ducks to cohabit. It is not much trouble to make a little duck-pond, or even a tub will do. I do not like to see stock ducks run about unless they have a place to go to wash themselves, as they always will dabble about in so much dirt if they can get at it. If they have a place to wash in, be it ever so small, they keep themselves nice and clean, especially when the water is changed in their little pond or tub occasionally. If pens of ducks are kept on grass runs they keep themselves quite clean if they have no water to swim in whatever. Those who have a good-sized duck-pond, and wish to keep several varieties, should divide their pond with wire netting, and allow it to go under the water at the least three feet if the pond is deep, otherwise when the ducks dive they may go underneath and come up the wrong side of the wire. Unless this precaution is taken in the breeding season they will get crossed. Whenever duck-ponds are made in the general garden, or where there is a fair space, a few drooping willows should be planted, with some large ferns of some kind, so as to droop round the pond to give a nice appearance.

A little house should be made near the pond, and it is well to have a covered run attached to it. A small one will do, even if it is only a few feet square, so that the ducks can run there, and be fed and watered before they go in the pond. Another reason why a run should be there is because very often when they are allowed to go in the pond early in the morning they will lay their eggs in the water; these are then lost. Ducks should never be let out before nine or ten in the morning to go to their pond unless they have laid beforehand. They should be kept from the pond and allowed to run in a meadow or where they can go ranging about looking for worms and insects, in which case they can be let out as soon as it is daylight, or before that even. Ducks, as a rule, lay from two up to ten in the morning, but the majority of them have laid by eight o'clock. Sometimes every duck will have laid by six in the morning. I have known as many as fifty eggs taken out of one pond where there have been only two ducks laying, and the owners have not known they were laying at all. As a rule, ducks are not so fond of going to lay in the nest as hens are. A nest of hay or straw should always be made for them, and a dummy egg put in after they have commenced laying. They will then usually lay in the nest, and frequently cover their eggs over after they have laid them. Those who are not accustomed to collect ducks' eggs must remember this, and always put their hand in the bottom of the nest to feel whether there are any there or not, because it is often impossible to see anything of them until they are uncovered.

Those who only want to bring up a few ducks for their own comsumption, say two or three broods during the twelve months, should buy the eggs, instead of keeping a pen of stock ducks just for this purpose, unless there is an opportunity of selling a few sittings to their neighbours or friends. That is to say, unless one has a fancy for keeping a few stock ducks, then that alters the case altogether. Some prefer ducks' eggs for breakfast—to such I should say, "Go in for a pen of ducks." Stock ducks are none the better for being kept too warm at night in their houses; as they appear to feel the cold very much the next morning. When they do feel the cold they will sit upon the ground, especially early in the morning when they are allowed to run out. They had far better be kept in an open shed than be kept too warm in a house. In all cases the houses should have good ventilation. I always have a window in my houses.

Where there is a flock of well-bred young ducklings there are usually some, when they get about six weeks old, which stand out from the others. That is to say, they are almost half as big again, and are developing a beak perhaps half-an-inch longer than others of the same age. To all appearances they look as though they were going to have an immense frame. These should always be picked out as soon as they begin to show signs that they will develop into large birds. They should be put in a nice roomy place by themselves and fed on good nutritious food. Very often people who go in for showing ducks go round looking at flocks of young ducks, and pick one or two out of one flock, and one or two out of another. I have known them give as much as thirty shillings each for them. These are the ducks which usually find their way into the show pens. I wish my readers to understand that, however large and good a pen of ducks may be, they will not throw all the offspring as good as themselves; there will sure to be a few among them much smaller than the parents, though some of them may surpass the old birds both in size and quality.

There are also freaks of Nature in ducks, just the same as there are in many other varieties of the feathered tribes. What I mean by that is, that the feathers in the wings will often turn the wrong way, instead of being close to the body they will often turn outwards. This of course looks very bad. When they come like this they should never be bred from. Some people do breed from them, but when they do the offspring look very unsightly indeed. Occasionally some of them will come with a little bunch of feathers on the top of their heads. If these are bred from again the little top-knots increase in size. I have seen a whole family of Aylesburies with quite a crest of feathers upon their heads. The feathers have been about two inches long, like those on a Houdan hen's head. I do not like to see them, as it looks unnatural in a duck. Of course there are people who fancy such ducks, and it is very easy to make a strain of stock ducks of this kind when a person who has a fancy for them has one or two sports come with feathers upon the head. Those who want to produce eggs in the winter should adhere strictly to the instructions given in the chapter on feeding. Give food as hot as the ducks can eat it, and if it is very cold weather give them a little warm water to drink. This may seem a ridiculous idea to some people, but if they get good laying strains of ducks and treat them in the way I have described in the chapter on feeding, namely, boil the corn and give it to the ducks quite hot, with the water in which it was boiled, and a feed of hot rice, with a little poultry powder occasionally, they will find it will well repay them for their trouble.

Ducks are now being kept merely for eggs and domestic purposes—that is, for one's own consumption—because a duck egg contains so much more nutriment than a hen's egg, and pastry-cooks always say that they go so much further in making pastry that they prefer two ducks' eggs to three hens' eggs. No one need despair in keeping a few ducks for laying purposes, if they have just a little place to keep them in, and can allow them to have a run sometimes, even if it is only up a garden path or passage, twice or three times a day.

Most gardeners are troubled with slugs, grubs, and wireworms in their kitchen gardens, so much so that, in some cases, the slugs strip everything off, viz., cabbage plants, peas, lettuces, &c. Where this is the case I recommend them to have a pen of ducks running about their kitchen garden just for, say, half-an-hour first thing in the morning, almost before it is daylight. The ducks eat up the slugs, grubs, and a great number of the worms; in fact, they will clear the whole garden of such pests. Occasionally they may peck up some of the young stuff which is growing; but if they are well supplied with green stuff in the little run where they are kept they will seldom do any damage in the kitchen garden. Ducks are quite different from fowls; they can find the slugs with their beaks very often when they cannot see them. They will collect them before it is daylight in the morning. I know some gentlemen who have large old-fashioned kitchen gardens who always keep a pen of ducks purposely to clear the slugs off. Ducks which have their liberty and run about (especially in the Spring) over a grass field or park scarcely require any feeding at all. I have known them not to eat a particle of corn for weeks when it has been given them; they have simply lived on the slugs and worms. But with those which are kept in confined runs it is quite another matter; they have to be fed by hand. Ducks are particularly useful to run over a ploughed field, as well as grass, as nothing clears the wireworms better than these birds. In one instance, I know a farmer who was troubled with wire-worms in a certain field, no matter what he planted it was always eaten up, but he bought 150 young ducks to turn into it and they cleared the field of the pests. They clear off slugs just in the same way. They will travel half a mile after worms and slugs, and if farmers were to adopt the plan of turning their stock ducks into the fields in the spring and autumn they would clear the ground of all kinds of vermin. Ducks will follow the plough the same as rooks do. Farmers find that ducks are among their best friends, particularly Indian Runners, and this hint may be taken by fruit-growers, especially where there is top fruit, such as apples, pears, and plums, as many grubs travel on the ground from tree to tree, and usually do this in the evening, and those who notice the ducks will find they go under the trees in search of these grubs after sunset.

STOCK DUCKS.

Management—Stock ducks and ducklings—Laying ducks—Meat diet for egg producing—Warm water for cold weather—The duck house: construction and litter, warmth and ventilation.

DUCKS are quite different to Fowls, and therefore they should be managed differently. Many people who keep a few stock ducks make a great mistake in the way they feed them after they have shed their feathers. No matter whether they are old stock ducks, or young birds which have never laid, when they have got well through their moult, most breeders feed them very liberally on rich food. This is all very well if they commence laying shortly after they have moulted, but old ducks usually moult from the middle of June to the middle of August; these are the two principal months for the old stock ducks to moult. The young ones moult when they are from 11 to 15 weeks old, and usually finish when they are about 16 or 18 weeks old. It is far better not to over feed them; if so, they become fat internally, which really prevents them from laying, and the owner has to

C

keep them for weeks and months idle. I have known young ducklings, hatched in March, not to lay until the following March; and I have known others, hatched in August, to be in full lay early in February; but of course the latter had different treatment. I have had ducks lay at six months old. The proper way to manage the stock ducks after they have moulted is to feed them very sparingly on ordinary food, such as bran, rice, and brewers' grains, mixed up with a little meal—viz., barley meal or sharps—and a few oats in their water, when they are put up at night, varying occasionally with a little maize. If ducks have a good range they should have nothing but grain, as they will find plenty of insect life until the end of October. They should be treated in this way until that date or even a little later with those who have not the convenience for breeding early young ducklings. There are three months when the stock birds should be kept in this way, and the cost of them should not be more than one penny per week each. When they have their liberty in running about the farm and ponds they want scarcely anything at all to eat; if you give them anything they seldom eat it. The exercise in a grass field or farmyard, where they have to work for their living, keeps their fat down. Ducks do not scratch in the same way that fowls do to exercise themselves. To make up for this a little grain should be thrown into the duck-pond when they are kept in confinement, and they will be standing upon their heads trying to fish this out; this keeps them busy for hours in the day. If the owner has only a garden path or passage for them to run up, they should be let out once or twice a day for a little exercise. When the owner wants eggs they should be fed on good nutritious food, when the eggs, as a rule, will begin to grow at once, and the bird's system is in a fit state to lay the eggs when they are ready. But when they are fed so well all along they are too fat for laying, and very often become egg-bound before they have laid ten eggs. When one begins to feed them on nutritious food to produce eggs, it should be given hot and in troughs.

I have had a large number of ducks sent to me for post-mortem examination just as they have commenced laying, and they have been lined with fat, so much so that the egg passage was blocked up. When the egg is passing down the oviduct something must give way. It is a very wrong thing to allow ducks to be hurried, especially when they first commence laying, as they are very full of eggs, and the thin skin which covers the eggs in the ovary is apt to break. If that does not give way, very often the egg gets broken by the sudden jerk while it is passing down the oviduct. A laying duck ought not to have to get up a high step or a bank. When they come out of the duck pond they should always have a ladder, or a plank, or something that they can walk up out of the water, without exerting themselves. Ducks have the power of holding their eggs, perhaps more so than any other variety of the feathered tribes. They have been known to lay three perfectly-shelled eggs in twenty-four hours. This is when they have been put out of their proper places when laying. Should such a bird be frightened in the least, it would either cause rupture of the egg-organs, or an egg would be broken in the oviduct. The latter means certain death, very often within twenty-four hours. When the owner or attendant wants to catch stock ducks he should never run them for that purpose, but merely drive them into their little house or shed, and catch them by the neck, not by the wing, as then they exert themselves very much more. When they are caught by the neck they should be carried under the left arm, holding the neck with the right hand. Those who are accustomed to hold or handle ducks always clip the body right across the back, holding the wings close to the body with the middle of the hand, so that the bird cannot get the wings loose to flap them, neither press its feet against anything. When they are carried in this way the exertion does not hurt them. When the owner begins to feed, especially for the production of eggs, he should begin to use a little meat, such as boiled greaves (granulated meat is better, only it is a little dearer than the greaves). The latter, too, is apt to stop up the passage leading into the gizzard, as there are pieces of bone in the greaves, and when little pieces of gristle and bone are attached to each other the duck sometimes swallows them, and they frequently block up the bird's gizzard, and it will die a very painful death. Pieces of meat skewers will often have the same effect. The greaves should always be boiled from an hour to an hour and a half or two hours— the longer the better; this will frequently soften some of the hard pieces. Whenever the greaves are used for ducks, the attendant should notice to see that there are no more hard pieces than can be helped which would be likely to do them mischief. But the granulated meat is all prepared and ground up fine. No harm can come to the ducks from it, as it is properly ground up, as it has to pass through sieves to prevent the large pieces getting through before it is sold to the public. The advantage of the granulated meat is that it does not require boiling, but merely boiling water poured over it, which saves time and labour, both the last named meat and greaves, when used, should be mixed up with the meal.

Lights and liver boiled, and also the intestines of a sheep or bullock cleaned and boiled are very good for ducks. This kind of thing should be put through an old sausage-machine, when a large number of ducks are kept. Where there are only a few kept an old worn-out carving-knife should be used—or any old long knife will do —to cut it up when it is placed on a meat-board. This can either be given alone or mixed with the meal; but I prefer the latter, unless one has a very large quantity of it. The intestines can be got at a very low price at a butcher's, or from a butcher's-man. They will almost give them away to get rid of them. Many people feed their ducks on horse-flesh; but I very much object to this system, as horses die from all manner of diseases, and when the flesh is bought up those who sell it are not able to guarantee that the horse or horses did not die of some infectious disease—viz., blood poisoning, which is frequently the cause, or poll-evil, or glanders. Horses die of any of these diseases, and when the meat is given to ducks it gives them diarrhœa very badly, and the eggs have a very nasty, strong taste—in fact, they are really uneatable. When duck breeders and keepers buy horseflesh they do not once give a thought as to what they are buying; if they did, I am sure they would not be so foolish as to purchase that which upsets their ducks and spoils their eggs for eating purposes. When a large flock of young ducklings are fed upon the diseased meat mentioned it gives the weaker ones diarrhœa so badly that many of them die within three or four days.

Those breeders who have several thousand young ducklings, and go in for buying horseflesh for food, usually buy the horse and have it killed on the spot; in such cases, of course, they know whether the horse is diseased or not. If it is a healthy horse I do not object to its being used for stock ducks as well as young ones. When horseflesh is used it should be cut up in small pieces, and thrown down to them. It is better not to mix it with the meal for the stock ducks, as it is very dry meat, and if it is mixed with the meal the ducks will frequently fetch it out and leave the latter in their troughs, or whatever they are fed in. In all cases where the meat is boiled for the ducks, the water in which it is boiled should be used for mixing up the meal, and in all cases during the winter it should be given them early in the morning, soon after daylight, as hot as they can eat it. They should be fed so they can clear every particle of it up. The attendant will soon know how much to mix, and when they leave any be sure and mix a smaller quantity next time. Of course, where there is only, say a drake and four ducks kept, it is well to use the refuse from the house for them, viz., scraps of cold meat, fat, vegetables, pudding, potato parings, &c. These should all be put into an old saucepan and boiled up together, then mixed with the meal. Ducks should be fed according to circumstances. There need not be a particle of anything wasted where there are ducks kept. Now we come to which are the best meals to use for the morning feed. This depends, of course, a great deal upon what is mixed with the meals. If there are a number of rich scraps, viz., meat, fat, &c., from the kitchen, then the meal does not require to be quite so nutritious. Sharps, which some people call thirds, with a little barley meal and biscuit meal, is a good thing. It is always well, whenever meal is used, to add a little biscuit meal, as it helps to mix so much better, and prevents the mixture from being sticky. The biscuit meal should be soaked in hot water, or a little put in with the hot scraps does just as well. Where there is but very little meat of any kind, it is well to have a mixture of meals, viz., a little bone meal, oatmeal, and French buckwheat meal, mixed up with sharps, middlings, or barley-meal. Those who are able to make their own meal will do well to make a little change for their ducks at times. Whatever is mixed in the way of meals should not be mixed too soft, but rather dry, just so that the meal clings together, and there are no little pieces of dry flour. If so, that is a great waste. If it is mixed too wet it sticks to the duck's beak, and they do not like it. I always find that the best plan is to feed ducks from a trough, unless one has a great deal of grass, and the food can be thrown down in a clean place on fresh ground every day, and even then the troughs are much the best when it is a very wet morning, as they prevent a good deal of waste. When the ducks are fed in troughs there need be no waste at all. If they do not quite finish up their meal, the troughs can be removed with their food in, so that they can finish it up at the next meal before they have anything else to eat. Green food is very necessary for stock ducks. Anything green can be cut up and given to them, but when cabbage leaves are thrown in to them the thick stems should be cut up in small strips, then the ducks are able to swallow every particle. Later on in the spring, lettuce is an excellent green food for them, in fact, nothing really comes amiss in the way of green food for ducks. Grass can be cut up short and thrown in to them. Where they have a pond it is a good thing to throw grass cut up short into the water, it keeps much longer than if thrown on the ground. It must be cut short; if not, it is apt to block the passage leading into the gizzard. When it is thrown into the water, and the ducks do not happen to eat it all up at once, it keeps nice and green, and they appear to enjoy it so much more when they skim it off the top of the pond. During the winter, when green food is very scarce and dear, it is well to boil up a few turnips and mangel wurzels, and mix in their food. These are a very fair substitute for green food. Of course, any time there are any small potatoes they can always be boiled up and given alone, or mixed in with the meal.

For the production of eggs, a little of the poultry powder advertised in the end of this book will be found an excellent thing, as in most cases it brings the birds on to lay. The powder not only helps in the production of eggs, but keeps the ducks in perfect health as well as assists the digestion of their food and helps to prevent them getting fat internally. In the breeding season, good wheat and French buckwheat I find are the two best grains to use for stock ducks, and I give barley and some good oats occasionally, but wheat and French buckwheat are my principal grains, and I always feed on the best I can buy; that is to say—good sound oats, weighing not less than 42 lbs. to the bushel; wheat, rarely less than 60 lbs. to the bushel; and French buckwheat, not less than 54 lbs. to the bushel, and the boldest barley I can get. In cold weather I like to boil grain two or three times a week for the ducks' evening meal. Whenever it is boiled, the water should be about three inches higher than the corn when it is put in, whether it is in a saucepan or copper. In this case, the corn does not soak up quite all the water. Give it to the ducks just as it is, water and corn together. It may be given warm, but not boiling. This helps the growth of the eggs very much. When the corn is given dry, or uncooked, for the evening meal, it should always be put in water. If the weather is cold, I like to put it in warm water; the ducks can eat it so much better, and enjoy it so much more when it is given in this way. The troughs come in very useful for this purpose. Grit for digestion should always be put in the troughs with their water or corn—the importance of the flint grit renders it necessary to have a chapter to itself. If ducks are fed in this way during the winter months, and the stock is from a good laying strain, the owner will most likely get a good supply of eggs. During the warm weather, in the spring and summer, they do not require so much care in feeding; they are sure to lay then if they have sufficient food to eat, even if it is not so good in quality. Twice a day is quite sufficient for laying or stock ducks to be fed. Feeding, as my readers will perhaps know, is one of the most important parts of duck-keeping; but it is not everything.

During the winter months management goes a long way. When it is a cold frosty morning the ducks should not be allowed out of their house, or covered run, whichever the case may be, until they have had a good warm breakfast. Ducks feel a cold, frosty morning perhaps more than any other variety of the feathered tribes, unless the ice is broken on the pond, so that they can swim about. When they can do this they do not appear to feel the cold so much; but it is well only just to let them out to feed and drink. Then drive them straight back into their house or run. During the very cold weather do not leave any water standing by them. In the middle of the day give them a little warm-water to drink, as they are usually very thirsty. This appears to warm their whole system, and they will flap their wings as though the sun was shining upon them. It is well to let them out for a little run, if it is convenient, after they have had the water. Some people like to feed the stock birds as late as six or seven o'clock at night. Ducks can eat in the dark as well as in the light, if the corn is put in a trough with the water. If they do not eat it all, the rats and mice cannot get at it when put in water, and there is no waste whatever, besides which the ducks enjoy it so much better; but mine are usually fed somewhere about four in the afternoon during the winter months. Ducks will lay more eggs if they are only allowed to run about a part of the day, and shut up the other part in their house or covered run. So, when people only keep two or three pens of ducks, one pond will really do for all of them, if they are let out in turns. Their beds, or sleeping places, should be kept very clean, and as dry as possible. Many people litter their house with straw, but peat moss is a very good thing, or fine shavings will do; farmers should use rough chaff. I like moss peat better than straw, only where it is used no water must be given inside the house; if so, it soon becomes wet and sour, and stains the white feathers of the ducks. Where it can be kept dry and free from the ducks' water, they do very well on it indeed, and keep much sweeter and cleaner than on straw. The peat moss is a great deal less trouble also. When the hot weather sets in, the stock ducks ought not to have much meat; if so, it causes them to shed their feathers too early, and they will frequently stop laying during the middle of June. They may have a good supply up to about the middle of May. Those who have prize Aylesbury ducks, which they intend for showing purposes, should avoid giving them maize, as that often gives a yellow cast to their plumage. It is well to give Pekins maize as it helps their plumage.

Ducks do not require any elaborate house. Those who

have pig-sties, like there are very often at farm-houses and at gentlemen's residences, will find these make splendid houses for, say a drake and four ducks, or two drakes and eight ducks. When the houses have to be built for ducks, they should be arranged in the same style as the pig-sties, viz., a place for them to sleep in, and a little outer yard cemented, so that the ducks can be fed outside their sleeping-place; then their house can be kept nice and dry inside. Another reason why they should have an outer pen is so that they can be closed up till they have all laid, as ducks are very often troublesome if they are allowed out early in the morning. For instance, they will drop their eggs about, and very often the pond is the receptacle for them. This makes it very awkward in the season when ducks' eggs are scarce and worth a good deal of money. My duck houses, where I keep two drakes and eight ducks, are about 4 feet by 6 feet, and the outer pen is about 4 feet by 12 feet, if one can spare a little more room so much the better. In all cases concrete should be used, or cement; if not, the rats will burrow holes into the house and roll the eggs into their holes. They can get at these better than hens' eggs, as the latter usually lay up in a higher nest, and at the same time do not lay till after daylight. Ducks do most of their laying early in the morning before daylight, and therefore the rats have an opportunity of clearing the eggs away. Of course, the ducks' houses can be made in height to suit the eye of the owner, but from 4 to 6 feet is really sufficient. At the same time it is very nice to have the houses high enough so that one can walk into them without stooping. In all cases have plenty of ventilation at the top of the house; ¾-inch match boarding, or feather-edged boards, is a good material for making the house. Where there are several breeds of ducks kept it is well to have the houses adjoining where there is a limited space, that is when a person makes a business of it : but where a gentleman has a large pond it is nice to have the houses dotted about a little way from each other, and a piece of wire netting run across the pond, so that the ducks may all be out together. A large box, such as a case that match boxes have been packed in, which can be bought at the oil shop or a piano case, makes a splendid little house for one drake and four ducks, if it is set a little on the slant, and a piece of felt nailed on the top, because these are really wood grooved together. A large barrel, or hogshead, could be made to answer the purpose. The match boxes can be bought from 2s. to 2s. 6d. each, and the hogsheads will not cost more than from 1s. 6d. to 2s. 6d. I just mention this because there are plenty of working men who may keep a small pen of ducks, and a little house may be made complete for about 3s. or 3s. 6d. Of course, this price would not include a concrete floor, neither would that be required, as the boxes are an inch thick, and the hogsheads or barrels are usually made with very hard wood, so that the old rats would have something to nibble at before they got through, and should they gnaw a hole through, it can easily be stopped up with a rag or piece of sacking dipped in gas-tar—if it is done in this way they will not take it out. Of course, those who have nice grounds, and like to see anything ornamental, can buy a duck-house for a drake and four ducks, or two drakes and eight ducks, for from 30s. to 50s., and they run up to £3 3s.; then a small run can easily be attached to the duck-house. If ducks are allowed to have their own way they like to sleep either on the pond or just on the edge of the pond. This does not hurt them in the least, because they are very hardy, but if they are housed, and kept from the cold winds and frosts, and are kept very dry in the house, either on peat moss, or clean straw, they will commence laying earlier in the winter. This is a most important thing. When the hot weather comes, it is always well to leave the door of the ducks' house open, so that they can either sleep in their little outer yard or inside. As soon as the very warm weather sets in, they will often begin to shed their feathers a month or two earlier than they ought, if they are kept too warm. While doing this, of course, they stop laying, which is a great loss to the owner. Jeyes' Disinfectant Powder or "Sanitas" are two of the finest things I have ever used. When there is a faint smell arising in the runs or houses it should be sprinkled about. Jeyes' Disinfectants or "Sanitas" are very valuable, as they help to prevent contagious diseases from spreading if put in the water. Should there be any smell arising from the duck ponds, a little of the liquid should be put in at once and it will take all the offensive smell away. No duck or poultry keeper should be without one of these disinfectants, as a preventive is better than a cure.

SITTING AND HATCHING.

Setting—Care of eggs during incubation—Duck eggs and incubators—The nest: formation and material to be used—Egg testing to ensure fertile eggs being sat upon.

SITTING and hatching is a very anxious time with most duck breeders, especially with the novice who has had but little to do with it. Many of them, perhaps, have not hatched out any before. With those who partly get their living by rearing young ducklings it is also a most anxious time. With them setting is seed time. Does not the same thing apply to almost everything? The gardener and the farmer, when they sow their seed, live in hopes that it will come up well, and then, when it is sprouting and looking healthy, the anxiety is not all over; the seed has to grow. Still, it is a great deal when the seed is up and the plant is beginning to shoot. That is a step in the right direction. Unless we sow we cannot reap. So it is with the ducklings. If we do not set and hatch we cannot rear, and unless we rear we have no fat ducklings to eat; therefore sitting and hatching is one of the most important parts of the whole branch of the business. It is important that the eggs should be properly set to ensure a good hatch. Now there are various modes of setting duck eggs. The majority of young ducklings are hatched under hens. Many years ago it was quite novel to see the hen bring up young ducklings, but that time has gone by. The hen used to be running round the pond in quite a way, seeing the ducklings swim, but as a rule they are taken away at once now. It is a waste of time entirely to allow her to run with the young ducklings. It is also a waste of time to let ducks sit on their own eggs, because any common hen will do just as well as a good bred-duck. Where one only has a few—say a drake and three ducks—and they lay their first batch of eggs and become broody, they should not be allowed to sit ; if so, it is a great waste of time. If they are shut off from the nest for a few days they will soon recommence laying. Duck eggs, as a rule, are worth more than hens' eggs, unless the latter are very choice. The only time ducks should be set is after the middle of June, or in July ; then they will shed their feathers while they are bringing the young ones up. In that case there is no loss of time, as they would have retired from laying eggs while they were shedding their feathers. There are many ducks' eggs hatched by incubators. Some breeders say that they do not hatch out well by this process, but my experience has been just the reverse. I always find that ducks' eggs hatch out better than hens' eggs in Hearson's incubators. I have tried them both together in the same drawer, and, as a rule find the ducklings hatch out quite as strong as they do under hens or ducks. Ducks' eggs require a great deal of moisture the last few days of incubation ; if they do not have this the inner skin becomes very dry. The best way to moisten them is to dip them in warm water, and then put them straight back in the incubator or nest again. When a hen or duck sits on the damp ground the eggs do not require damping at all, as the warmth from the hen or duck gradually draws the moisture out of the ground or soil. When they are set in this way they usually hatch out well, and are but little trouble. It is well to make the nests with damp soil, beaten down hard, with a little fine hay at the bottom, rubbed up very short, as the shells of ducks' eggs are usually very thin; therefore the nests should be made so that the eggs do not bind together when the hen or duck goes in on them. If the nests are made fully large enough, with the short hay at the bottom, the eggs turn over so much easier, and do not crack so quickly. The hen or duck usually turns them over twice during the twenty-four hours, sometimes more. It is very essential therefore that the nest is made so that the eggs can roll easily. Some people set ducks' eggs on the solid ground, without anything at the bottom of the nest, but this is wrong. If there is nothing but the earth on the bottom of the nest, when the eggs chip and the beak of the duckling comes through the shell on to the hard ground, out of eleven eggs there may be two or three turn so that the beak is on the ground, and when this is the case the duckling will often get suffocated. But should there be a little hay, or something loose or porous at the bottom of the nest, they are able to breathe, and do not get suffocated at all. It is possible to set the eggs on the bare ground and have every one hatch out without losing a single duckling. At the same time there is a possibility of their getting suffocated, so it is far better to have the nest properly made. I have known from two to five suffocated in the nest through having nothing loose at the bottom. Ducks in their wild state always provide something for the bottom of their nest. A fair-sized hen can cover 11 duck eggs, and a good large one 13. The eggs have a very clear shell, whether they are white or blue. This enables them to be tested at an early date to see whether they are fertile or not. An experienced hand can tell whether they are good or not at the second or third day, but those who have not had experience in testing them should leave them seven or eight days at least before they test them. It is simply a waste of time to allow a hen or duck to sit on the eggs for a month and not test them to see whether they are fertile or not, which many people do, because if they are not fertile they can be taken away at the seventh or eighth day without running any risk, and others can be put in their place. Should the owner not have a sitting of ducks' eggs ready, they can put hens' eggs under instead, especially where there are a large number of ducks' eggs set at one time, and every egg is dated, so that the eggs can be removed from one hen to another. For instance, suppose four hens are sitting at one time, and the eggs are all tested on the seventh or eighth day, all the unfertile ones should be removed and saved to boil up for the ducklings. Those which are fertile should be removed from under the hens till the latter have a full sitting each, fresh eggs being put under in the place of those which have been removed. If a hen has six fertile eggs left, it does not do to put five more cold ones under with these six; if so, the cold ones are liable to chill the warm ones, as it takes 24 hours before they are warmed through. Suppose there is a hen with only a few eggs, say from four to six fertile, and there are no other hens to take the eggs. In that case put some bran in the oven to get well warmed through. When warmed through, put the bran in a box, and put the eggs in the warm bran, covering them well up with the latter, and wrap the box up in a piece of flannel; then set it near the fire, or on the hob, so that it does not get too hot. When done in this way the eggs will keep warm for from ten to twelve hours, and it will not injure them in the least. While the eggs are in the box put some fresh ones under the hen, and if she sits on them for 10 or 12 hours they will get warmed fairly well—not warmed right through, but just enough so as not to injure those which already have young ducklings in them—so that the former can be replaced under the hens together. Do not forget to put the date on the large end of the egg in ink; then when the first batch of eggs hatch they can be removed. The young ducklings should be taken away carefully, so that they do not squeak. A small box should be provided from 3 in. to 4 in. deep, and a sheet of cotton wool placed at the bottom of it. The box should then be put by the fire to warm. The ducklings can then be placed in the box and covered over with another sheet of cotton wool. It is the best to let them remain in the box from 12 to 15 hours, then they become strong. If the young ducklings are taken away very carefully, 19 hens out of every 20 will hatch the remaining eggs the following week. It is always well when ducks' eggs are hatching to look at them a few times, as the eggs are often chipped from 14 to 26 hours before they come out. There are two skins inside the egg, the inner membrane, which is the skin next the duckling, becomes dry and adheres to the fluff tightly, so that it has no power to struggle and break the shell open. When this skin is noticed to be dry it is best to take a penknife and just separate the skin from the fluff very gently. Do not go into it too deep; if so it is liable to let the blood out, as this is the last thing they take up. In no case should the eggs be shaken in the least. Many people when they have young ducklings hatching out take hold of the eggs and give them a gentle shake to test whether they are good or bad. By that sudden shake they often rupture the blood vessels, and usually it will kill 17 out of 20, and sometimes every one of the ducklings, unless they are ready to come out, then it does not injure them. When the time is up for the eggs to hatch out and there is no sign of life, hold each egg in the left hand firmly, and make a little incision in the large end, which should be held away from the operator, because, should the egg turn out to be rotten, it will occasionally go off like a gun, and it is not very pleasant if the liquid flies in one's face. Open a small hole, about the size of a threepenny piece, and if the duckling is alive the skin will be quite white inside the egg;

if the duckling is dead the skin begins to turn dark in the course of two hours. In many cases ducklings cannot break their shells to liberate themselves. If the beak lies a little on the slant, that is, too near either end of the egg, as this is oval, when they peck at the shell their beak slides, and if the shell is not opened they die in that state. Some duck breeders will have noticed that when they have opened the eggs they have often found ducklings fully developed in the shells, but quite dead, and they wonder what can be the cause; but what we have stated is the principal cause for this. When a small hole has been made in the egg it is well to turn it round to ascertain where the beak lies; it can usually be seen moving about an inch, or an inch and a quarter from the large end of the egg. If the operator can find the position of the beak, he should draw a pencil line from the incision which is made to where he thinks the beak is, then put the thumb across the pencil mark and hold it down firmly, make a small incision with a penknife just where he fancies the beak is. If the egg is not held firmly, it will crack from the hole that was first made in the end, which often bleeds the duckling to death, so great care should be taken. This is one advantage of having an incubator, as all cracked eggs can be put in and they will hatch out. Duck eggs take one week longer to hatch than hens' eggs, namely, 28 days. Occasionally one may hatch out on the 26th day, and frequently they will come out on the 27th day. Muscovies take 34 or 35 days to hatch out. This is a most remarkable thing. When the Muscovy drake is crossed with the

Aylesbury or any other breed they will often come out two days over their usual time, viz., 29 or 30 days. Where duck breeders hatch out a large number of young ducklings it is well to have an incubator, even if they do most of their hatching with hens, as this is so very useful, even if it is only a small one, to put the eggs in which may be a little late in hatching, as well as some of the cracked ones. The latter are just as good for hatching purposes, if there is no weight put on them, as those which are not cracked at all, so long as the inner skin is not broken. I thoroughly recommend an incubator where people hatch a large number of ducklings, as it brings about such a saving of time. I have never found any incubator hatch ducks' eggs so well as Hearson's. I run my incubator at 104 deg. for ducks' eggs during the first fortnight, and then down to 102 deg. for the rest of the time. I find this does better than keeping it at 104 and 105 deg. all the time, as many do. By following this plan a less number die in the shell, and the ducklings are much stronger when they are hatched out.

FEEDING AND MANAGEMENT OF YOUNG DUCKS.

Rearing ducklings—Feeding young stock—Stock and exhibition ducks—Ducks for home consumption—Ducks for market—Aylesbury duck management—Duck houses and barns—Fattening—Warmth for growth.

AFTER the young ducklings are hatched, the next thing is to bring them up; but they are less trouble to rear than any other class of the feathered tribes I know of, young geese excepted. After the first three or four days are over they are practically no trouble at all, if they are fed, watered, and kept clean. But, say some of my readers, "Which is the best way to do this, so that we may produce fine young ducklings for the table, weighing at least $5\frac{1}{2}$ lbs., up to $8\frac{1}{2}$ lbs., at about ten weeks old?" Of course there is a great deal in feeding, as regards getting size in the young stock, but it is impossible to get size unless the eggs have come from proper stock ducks; there is little difficulty if the eggs originate from good-sized, healthy stock.

When ducklings are hatched by hens, some duck rearers say it is best to let them remain under the hens from 12 to 18 hours after they are hatched, unless one or two happen to come out considerably earlier than the others; these should be removed. Perhaps I break away from the general rule here. Some people leave them with the hen, but I always take them away and place them in a box, and cover them well over with cotton wool, which I warm well through first, then I wrap the box up in a piece of flannel and put it near the fire, taking care that it is not near enough to get very hot. The reason why I take them away is, so that the hen shall not step upon any of them and kill them. They often do this when the young ones are just hatched out, because frequently the hen gets restless. In some cases I have known her to be frightened at them after they come out of the shell, but this is very rare. When hatched by an incubator I leave them in the drying box from 15 to 20 hours after they are hatched, then they become quite strong. Where one has an incubator, and the ducklings have been hatched by hens, it is best to remove the young ones from the hens into the drying-box of the incubator. If this is done it saves many of the ducklings from being killed by the hen stepping on them and crushing them in the nest directly they are hatched; as this will often happen when they are not seen to. Of course it all depends upon circumstances. Many people, when they have a brood of young ducklings hatch out, become very anxious about feeding them as soon as they can stand upon their legs. This is quite unnecessary,

Nature provides for all the feathered tribe until they become strong enough either to feed themselves or to be fed. Young ducklings are strong enough to feed themselves after they have been out of the shell about 15 or 18 hours. They really do not hurt without food even from 24 to 30 hours after they are hatched out. When they can run about nicely, I like to put a little food down to them. My first feed, as a rule, is a little water in a saucer or some little shallow vessel, and some coarse oatmeal sprinkled on the top of the water. While sprinkling the oatmeal on the water, it is well to let some fall on the ducklings' backs; this will teach them to peck, as when they see it on each others' backs they peck it off and so get the taste of the food. If they do not appear to begin to eat readily, one or two should just have their beaks put in the water; then when they get the taste of the oatmeal which is floating on the water, they commence to swallow it at once. Then either stale bread crumbs or oatmeal should be mixed with hard-boiled egg. If it is early in the season they should have hard-boiled egg for several days. When they are required to make large stock ducks for exhibition, or if one wants to bring them up to an extra size, they should have egg give them till they are from 7 to 10 days old. But in the ordinary way, if they have egg given them till they are two or three days old that is quite sufficient in the cold weather, or early in the season. Biscuit-meal just soaked in hot water, not too much, just enough so as not to make it sloppy, but nice and crumbly, should be given them; then they can peck it up nicely and enjoy it very much. It is little trouble, and in the end comes as cheap as anything which can be given them, as there is no waste. Many of the Aylesbury people for the first few days toast a stale piece of bread before the fire until it is burnt quite black, then soak it in cold water and put it in the water vessel; the young ducklings will then nibble away at it.

The people in the duck counties consider this a very fine thing. No doubt it does them no harm, and they think it does them a deal of good, which it may do. I have tried young ducklings with it, and some without it, and I do not find there is any difference. I prefer giving oatmeal in the place of the burnt bread. The former is less trouble, and the youngsters seem to enjoy it quite as much, and at the same time it is very nutritious. Those who are interested in rearing young ducklings will do well to give both systems a trial for the first fortnight. Of later years I have used nothing else but the oatmeal and biscuit meal for that period, but after that their appetites sharpen up and they begin to eat well. Then I give them a mixture of my own, and a little granulated meat, finely ground, putting a little biscuit meal in to prevent it becoming sticky. Meal in all cases should be mixed rather dry and crumbling, not too wet. It is a very nutritious meal, and the young ducklings grow wonderfully fast on it. I have had them weigh $6\frac{1}{2}$ lbs. at eight weeks old, and $8\frac{1}{2}$ lbs. at ten weeks old. If they are fed in this way till they get about five weeks old a good foundation is laid, and they make immense frames. When it is early in the season, and young ducklings are fetching a long price, this food should be used till they are about six or seven weeks old. Then add extra meat and fat, and also a little barleymeal to the other things. This is what I term the best way of feeding young ducklings when one wants to make splendid birds either for stock or fattening. When the ducks are only going to be kept just for ordinary stock the extra barleymeal and fat should be kept out, and a little extra oatmeal and bonemeal added.

There are plenty of people who only rear one or two broods at a time—just for their own family consumption—who would say it was too much trouble to mix all these things. It is possible to get very fair ducks and not use all these different kinds of meal I have described. Numbers of people rear very fair ducks, and give them nothing else but a little barley-meal and sharps, mixed together with some granulated meat or scraps from the house, and they do not have anything else from the time they are four days old. But these do not grow to anything like the size of those fed in the manner I have mentioned. When young ducklings are intended for selling for stock, after they are about five or six weeks old they should have a little grain put in their water troughs the last thing at night, such as French buck-wheat, wheat, or barley; if not, the digestive organs become weak. When they are about 12 or 13 weeks old some of them begin to shed their first feathers. After this they should have good oats put in their water at night. This will keep their plumage in as good condition as anything they can have. A little sulphur put in their soft food in the morning, say a teaspoonful to every six or eight ducklings, about two or three times a week, will help the feathering very much. The first week young ducklings should be fed about every hour-and-half in the day, and late in the evening for the last meal. The second week feeding them about five times a day, or six, if one has time, will not hurt them. The third week four times will be sufficient. After they are four weeks old I usually feed mine three times a day, although many people feed them only twice, but they should really be fed three times until they are full grown. In all cases feed them so that they will clear up the food properly, and put their drinking vessels a distance from the troughs which they feed out of; if not, there will be a great waste of their food, as they will run to the water with the meal in their beaks just the same as the old stock ducks do, and this causes a great waste.

When the young ducklings are required for killing purposes, after they are five weeks old they should not have water more than twice a day—the latter should never be allowed to stand by them all the time. Too much water is very injurious to them. It is well to begin feeding the ducklings before the water vessels are put down, or else some of the young ones will drink instead of eat, which often brings on diarrhœa and giddiness, so that they cannot stand upon their legs. Many people would rear ducklings for their own table use, only they have an idea that they cannot be kept properly unless there is a pond or stream of water for them to swim in. This is a great mistake. All early ducklings should be kept very dry; they grow so much faster, and it prevents cramp in them a great deal.

If the young ducklings are let go into the pond, or stream, it will often bring on cramp or weakness of the loins. Even later in the season, say the middle of April, if they are allowed to run in a pond at all, just to wash, they should not be allowed to remain in the water for more than ten minutes. The Aylesbury people allow them to run in a stream or pond about twice a day, to wash themselves, after the 1st of May, but I do not allow my young ducklings to go into the water until they get all their plumage. I always think they grow so much faster when they are not allowed to go into the water at all till they are about ten weeks old.

As I said in the commencement of this chapter, rearing young ducklings is an easy thing, and they may be kept thicker on the ground than any other class of the feathered tribe. People often wonder how Buckinghamshire and Bedfordshire duck breeders bring up so many ducks; they think they must have large farms, but it is nothing of the kind. There will often be found five hundred ducklings on less than half an acre of ground, and earlier in the season, before they are let out on grass, considerably thicker than that. I will give the Aylesbury breeders' system of managing young ducklings as regards room and accommodation. I have seen what they call a duck barn; these are made with walls of plaster (which I believe is white clay), and most of them have thatched roofs; just a few of them are boarded buildings, but very few. Almost all of the sheds or barns which they are kept in are thatched very thickly, which keeps the ducklings nice and warm in the winter and cool in the summer. It prevents the hot rays of the sun from penetrating, and the sharp frosts in the winter from reaching the insides of the sheds. One duck house I measured at a village close to Aylesbury was 12 feet square, and it had 250 young ducklings in it, ranging from two days to four weeks old, and every one of them perfectly healthy, but I ought to mention they were divided into four separate lots by thin boards for partitions. Those who have not seen them would scarcely credit that young ducklings could thrive and grow where they were kept so very close together, having no hens running with them. They are kept clean, being bedded down with fresh wheat or barley straw every morning. They are let out of these places, where they are so very thick together, to feed and water in what the people call an outer yard, so that they have more space while they are feeding and drinking. After they are five weeks old the Aylesbury people very seldom feed them more than twice a day; some breeders feed them three times. They usually have as much as they can eat at a time, but nothing is left to stand by them, neither is the water allowed to remain near them, as they consider too much water very injurious to the growth of the ducklings. The Aylesbury people have always been brought up to this kind of management from children, as have also their fathers, their grandfathers, and their great-grandfathers before them. They feed the ducklings in long troughs generally. They also use a quantity of greaves, or granulated meat, in their food. Those breeders who rear a number of young ducklings have a copper boiling with some kind of meat or refuse in it every day to mix in with their meal. In one barn I saw, twelve feet by fifteen feet, there were one hundred and forty ducklings, almost full grown, which had never been out more than from five to ten minutes at a time since they had been hatched. Every duckling was perfectly healthy and growing fast. I usually visit the duck counties once a year. I have done so for many years, driving round to different villages, where I find from fifty to five hundred, even at small cottages. They are only considered very small breeders if they have not as many as five hundred or one thousand just in the height of the season. I visited one district in one day, which had from forty to fifty thousand young ducklings in it. All of these find their way into the London markets.

The sheds, or barns, where they are kept, are divided off with either movable partitions or thatched hurdles, so that there are from 25 to 50 in one lot. It would not do to put them all together; if so, they might tread upon each other, unless they were all about one size. The Aylesbury people are very careful after they are fed and watered to let them lie quiet. They do not go to disturb them between meals, and they keep them rather dark, as they think they grow faster. I use peat moss for the young ducklings instead of straw, as straw is very dear near London. I find they do equally as well on the peat moss as on the straw, but they must not be fed or watered upon it; it must be kept perfectly dry. It saves a great deal of labour, because the sheds do not require clearing out every day when the peat is used. Every time they feed it should just be raked or forked over. None of the excrement need be taken up, as peat is a disinfectant, and, being very dry and of a spongy nature, the excrement from the ducklings adheres to it, therefore there is no smell attached to it. If the peat is put about 2 inches or 3 inches thick it will last a good time; of course it depends upon how many ducklings there are in the house, and what age they may be. If there are, say, one hundred birds in a 12-feet square place, divided off, three lots of peat moss will do well for about ten weeks, so that it amounts really to only clearing the house out three times in ten weeks, which means a great saving of labour. Should the peat moss become wet at all, remove the whole of it at once. Of course, when they are kept as thick as the breeders in the duck counties keep them, the moss peat would want changing oftener, according to the number of ducks and the circumstances under which they are kept. While it is perfectly dry and there is no smell arising from it, there is no necessity to change it, only it should be forked over every day. Some of the Aylesbury people in the very hot weather change their ducklings, that is to say, they have other pens out of doors, with thatched hurdle to form the sides and ends, or some rails with straw between, so that no draught can get to the ducklings. The railing is about 3 feet high, and the top of the little shed is about 5 feet high, so that there is about 2 or 3 feet where the pure air can get to them all round without causing a draught. Where straw is dear and scarce, the same kind of thing can be arranged with canvas at the sides and top. This will be more fully explained in the chapter on building houses and sheds. Young ducklings should never, under any consideration, be left out in the hot sun in the middle of the day. There are many people lose their young ducklings in the hot weather. Frequently, where there are large numbers reared, there will be from four to ten die in a day. Sometimes, where only a few are brought up, two or three die every day; that is, of course, when they are allowed to run in the sun in the hot weather. I have a large number sent me for post-mortem examination where every organ of the body is perfectly healthy. A young duckling's skull appears to be so thin that the sun affects the brain, and often causes death in a few minutes. Some of them will stagger about for an hour or two before they die, but generally they are found on their backs. It is well to use a good quantity of boiled rice in the hot weather with the other meals, as it is very cooling. A meal entirely of rice should be given occasionally, but it must not be made up too sloppy. Put it in a cloth to boil it, but take care that the cloth is tied loosely; if not, when the rice begins to swell it will burst the cloth. Wherever there is skim milk or buttermilk, such as there often is at a gentleman's residence or a farmhouse, always use that to mix up their meal with, when it can be spared.

Those who only rear ducklings for their own consumption, or rather, rear them so that they have some to kill the whole of the spring and summer, should have a few small places made just about four feet square, so that they can be put in at different ages. A place from four to five feet square does splendidly for from six to ten ducklings, and gives them plenty of room. In all cases they should be kept clean and dry in their sleeping houses. I have made no mention of green food for the ducklings. Many breeders do not use any of this at all, or at any rate very little. I prefer using it when I can get it, as it is a very fine thing for the health of the young ducklings. Lettuce, cabbage leaves, cut up fine, grass, tares, or clover, run through a chaff-cutting machine, where there are a lot of young ducklings reared, are all capital things for them. When they have a good supply of green food, they do not require so much other food. When I am fattening young ducklings, I like to give them a little fattening powder, which is specially prepared for fattening ducks and fowls, about five times during the last fourteen days of their existence. Since the introduction of the book on ducks, there have been many thousands of people rear ducklings for their own domestic purposes who have never thought of such a thing before, having an idea young ducklings cannot be reared without a stream or pond to swim in. For instance, those who have a large family and can consume a great deal of this kind of food, when they are eating roast duck, be it ever such a luxury, they are keeping the butcher's bill down. Such people ought to have broods coming along, one after the other, from March to August, so that they are kept well supplied all through the season. Supposing they are fortunate, and the eggs hatch out well, there may be a few young ducklings to spare; then a present of a nice young fat duck is very acceptable; otherwise there are neighbours who are only too pleased to buy them. It is far better to dispose of them this way, where there are a number ready at one time, as after they reach the age of eleven or twelve weeks they begin to go back in condition.

Some of my readers may ask, "How do farmers rear young ducklings when they are always allowed to go into the pond from the time they are hatched, when you say it is injurious for ducklings to go into the water?" Now it must be understood that it is very seldom farmers hatch out any till the end of April, or the beginning of May, when of course the water is not so likely to give them cramp. If they were to hatch out in January, February, and March, and then let them have their fling, the farmer would most probably find the death-rate would be very high. Where people have well-bred ducklings, especially if they are bred from good stock, they will find a few of them appear to stand right out from the others, with larger frames and beaks. These should be picked out of the flock when they are from five to seven weeks old, put in a place by themselves, and fed as I have described. They should also be kept much thinner on the ground than those which are being prepared for the market or table. Unless ducklings promise to be good stock birds and likely to make a fair price, they should always be sent to the market before they are twelve weeks old, or else killed and consumed by the owners. I should advise those who have no warm house for the young ducklings when they are hatched, and especially those who have only a few ducklings, say one or two broods, to let the hen run with them, if it is cold weather. They seem to grow so much faster if they are kept fairly warm. Where there are a large number in a shed together they keep each other warm, unless the weather is very severe. When the warm weather sets in they do not require the hen or foster-mother. Some people hatch out in January, February, and March, and put an oil stove in the house for the first two weeks without a hen or foster-mother.

HOUSING YOUNG DUCKLINGS.

Houses for ducklings—Penning in open air—Ducks in orchards—Windows and ventilation for duck houses.

HOUSING the young ducklings is an important matter. When they are hatched early in the winter, they want keeping warm, whether they are intended for the markets or for stock ducks; if so, they grow much faster. When I say warm, I do not mean having a lot of artificial heat, but that they want nice warm snug houses, and these should be arranged so that the bottoms of them are perfectly dry. They should have concrete or brick floors, which should in all cases be covered with peat moss, or nice clean straw. Of course, the size of the houses must be arranged according to the number of ducklings. When they are intended for killing purposes they can be kept very much thicker on the ground than when they are being reared for exhibition purposes. In the duck counties they have them very thick. In a house twelve feet square they will frequently have over 200 young ducklings; they consider they keep each other very warm when there are plenty of them. There is, no doubt, something in this, but they should always have plenty of ventilation above them. Where one breeds many young ducklings, it is well to have good-sized houses, so that the ducklings can be divided off into separate lots in each house, according to their size. The partitions should be movable, and the easiest way is to drop them into sockets, that is, nail a piece of small quartering to the boards or wall and drop the partitions in, as they have to be removed every time the ducklings are let out to feed and water. When the houses have been made specially for the purpose, then partitions should be made similar to stalls in a stable, so that the ducklings can be parted off, but the partitions need not be more than eighteen inches high. However the building may be constructed, in the cold weather it is well to pad it round with wheat or barley straw; this will keep them very warm. I do not like artificial heat, unless it is a severe frost, and there are some early young ducklings; then the paraffin oil stoves are the best things to use. But these are not required unless it is a very severe frost. I do not recommend ducklings to be kept quite as thick as our duck-breeders usually keep them, especially when the weather gets warm. I think a house twelve feet square is much better for 130 ducklings than 250. In all cases they should have plenty of ventilation, and the duck-houses should be arranged so that there is a place to allow the young ones to come out to feed and water. They should only be allowed to feed and water in their sleeping places the first week or ten days. It is well to have an open shed to feed them in, in case it rains. In fine weather a good run does them good when they come out to feed and water. Of course, in all cases they should have a run to stretch their legs. During the summer time it is well to make them little places outside, so that they can have a run and get fresh air, at the same time they are sheltered from the burning sun. The place can be arranged according to the number of ducks. It is best to have square places, with thatched hurdles and four posts. Drive a post in each corner, so that the hurdles can be fastened to the posts. The latter should be about 18 inches longer than the hurdles are high. In this case a thatched hurdle can be laid on the top for a roof; this keeps them from the wet and sun also. Where straw is scarce little places can be made of canvas or thin boards; if it is not convenient to get canvas, cut some old sacks or bags open, and stretch them out on a frame, then tar them both sides. If they are tarred the second time, as a rule it will keep all the wet out. Where straw can be used it is better, as the places keep cooler. An orchard, where there are good large trees, is a very fine place to bring young ducklings up, as they have good protection from the sun. When one has had a little experience he can soon tell what is best for them as regards their management. Experience is the best teacher, only some of us have to pay very dearly for it. I have not laid down any hard-and-fast rules as to what the houses are to be made like. It does not matter much what shape they are, so long as the ducklings are not overcrowded, and have plenty of ventilation; it entirely depends upon circumstances. Duck-breeders must, to a certain extent, be ruled by their own judgment. As far as possible the houses should be facing south; in all cases the windows should be south. Some people do not like windows in duck-houses at all. I always like some kind of window, even if it be ever so small. I am fond of light for everything. If the window is large, and the weather very bright and sunny, it is an easy matter to put a curtain over it; then in dull weather a large window is a boon. As I have said before, things ought to be arranged according to circumstances; but in any case I never think of building either a fowl-house or duck-house without a window.

PURE-BRED DUCKS.

AYLESBURY STOCK DUCKS.

Aylesbury ducks—Points of the Aylesbury duck eggs *versus* chickens' eggs—Stock Aylesbury ducks and trickery—Good breeders for good ducks—Pens of Aylesbury Ducks: how to select and start—Rouens: their characteristics and crosses—Pekins: their points and peculiarities—Cayugas: their antecedents and merits—Muscovies: their habits and heaviness—Black East Indian, "little and good," the points and perfections.

THE celebrated Aylesbury ducks originated from Aylesbury, in Buckinghamshire, and they are a very old breed. I have never found a trace of how long the Buckinghamshire people have been producing this race of ducks, but have heard several times from old people that they have been popular in Buckinghamshire for nearly two centuries. Aylesbury ducks appear to have taken the place of game in the London markets—that is to say, after the game have ceased the young ducklings come in season. No doubt this is why Aylesburies have made such a name in England. They have been a luxury for the rich for at least two centuries, but I cannot say how much further back than that. I have not found any record of the time in any old poultry book. Aylesbury ducks are now in every part of the world where the white man's face shines, and were considered for many years the best table duck which could be produced. In most cases they realised the most money in the London markets. Not so now. Indian Runners will fetch as much as pure Aylesburies, and so will Indian Runner Aylesbury first-cross ducks. In some cases the Pekin Aylesbury cross will fetch, perhaps, a trifle more because they are rather larger than pure Aylesburies. There are many hundreds of ducks sent to the London markets now, particularly cross-breds. A few years ago there would be from 800 to 1000 early Aylesbury ducklings sent to the markets to about 20 of any other breeds, but things have altered in this respect. I am often asked by those who keep ducks, as well as by some who intend keeping them, "Which are the best breeds to keep as pure?" My answer to them is Aylesburies and Indian Runners, but the former are the best table birds, as they are bigger than the Indian Runners, they are some of the finest table ducks which can be produced in a pure state. As I have said, they are good layers of fine eggs. Pekins will sometimes lay very early in the season, but I have often noticed my Aylesburies have commenced laying earlier than the Pekins, when the treatment of both breeds has been similar. There are many people call Pekin ducks Aylesburies. Some who have kept ducks for years do not know the difference between the two. Aylesburies should have pure white plumage, with a flesh-coloured beak and orange legs. Pekins should have bright yellow beaks, and their appearance is altogether different from Aylesburies. They are a much shorter duck, and stand up very erect; more the shape of a goose. Their breast and head are straight up, and also their tail. Their plumage is of a yellow cast, as is also their flesh, while the Aylesburies are quite straight in the back, and they also carry their tails straight out. The breast, which usually goes by the name of "keel," should almost touch the ground in a good-bred Aylesbury the second year. When a good Aylesbury and good Pekin are put together side by side, there is no comparison between them, as will be seen in the drawing in this little work. Young Aylesbury ducklings are quick growers, fatten easily, and give but little trouble. To any one who thought of going in for a pen of pure ducks, either for their eggs or to produce table birds, Aylesburies are the ducks for them. They vary in their laying qualities. I have known them not to lay more than 50 eggs in the twelve months, and I have had them lay 150 eggs in that time. Of course it is an exception for them to lay that number, but they will do so occasionally. Some people like having ducks' eggs for breakfast in preference to hens' eggs. In the first place they are much larger, and some people think they are far more nutritious. A pastrycook would prefer three ducks' eggs to five hens' eggs. A duck's egg seldom weighs less than two ounces and a half, and they will often turn the scale at three ounces. I consider a pen of good laying ducks are quite as profitable to keep as a pen of good laying hens; in fact, they can be kept where hens cannot. Stock ducks eat but very little when they are full grown and are not laying. I have often noticed they have not consumed so much food as the hens. If they have a good range of a meadow, and such a place as a farmyard in the country, with a pond or brook, when they are not laying they seldom want food at all; they get their own living. I would warn purchasers of buying Aylesbury stock ducks of ordinary advertisers. Go to well-known breeders. There are good duck-breeders who sell first-class reliable stock. Some people give from 7s. 6d. to 21s., and others go as high as £3 3s. for good-bred Aylesburies. Many of them are sold for Aylesburies, but are only really cross-bred. This is a disappointment, and it may be months before the owner finds out that they are not pure. I have known people who have kept ducks for years, who have offered me their stock of young Aylesburies, and in some cases I have asked them what colour their beaks were, and the answer has been, "Almost every bill is as yellow as a guinea." Of course they have been Pekins, and not Aylesburies at all. In one or two other cases I have asked them to send the ducks on approval, especially where I knew the people and thought they would know what Aylesbury ducks were; but I have had to feed them and send them back at once. Where people do not know any better, ignorance is bliss. Those who take the trouble to go through this little book will be able to judge for themselves whether they get a cross or pure stock. There is one thing in Aylesburies which I should mention. If they are allowed to run on grass it gives their plumage a yellow cast, and particularly the old stock ducks when they are shedding their feathers. If the young ones, or even the old stock birds, are allowed to run about in the hot sun, their beaks become a yellowish tint, but this does not alter the shape of the duck, only the colour of the bill. Some of the bills will get much more tinted than others. I should recommend those who are breeding for purity to buy ducks with good coloured bills. Of course it does not make any difference to those who are going to breed for table purposes, and not for exhibition at all. When the bills are tinged by the sun the ducks are considerably cheaper, even if they are very fine birds. So the purchaser here may gain a benefit by the ducks being reared in the sun; only be sure they are pure if they are bought as such. When one is starting a pen of Aylesbury ducks, see that they are unrelated, and have them from a good reliable breeder. If there is any question about this, have the ducks from one breeder and the drake from another. I do not mean to say that there are no duck-breeders who have got unrelated birds. Those who keep a number of pens have generally got them unrelated; but it is best to be on the right side. It is never well to breed from brothers and sisters, because it oftens brings the young ducklings out weakly, the eggs are not so fertile, and many of them become cripples. Stock ducks for ordinary breeding purposes—that is to say, breeding for table—should run from 7s. 6d. to 12s. 6d. each, and a good drake from 10s. 6d. to 21s. It is always better to spend a little extra money on the drake, because every duckling is related to him. It does not matter if the ducks are a little smaller for breeding from, if the drake is a good-sized one. When Aylesbury ducks are bred to such a size they are seldom good layers. It is not often one finds a large duck an extraordinary layer, therefore it is better to get the size from the drake. Those wishing to go in for expensive Aylesbury ducks I should recommend to pay from £4 to £10, according to quality, for a good pen of Aylesburies. Three good ducks and one good drake from the best blood ought to be bought for about £5 5s. There are, of course, special ones worth considerably more, but it is only those who are experienced who should be tempted to give more than £5 5s. for a pen of good ducks. Sometimes they will run up to £10 each when they are really extraordinary, but I do not recommend the novice to pay such a long price for what he does not thoroughly understand; if he does he will find his balance sheet on the wrong side at the end of the year. A good-bred Aylesbury duck should have a thin neck and a long head and bill, and the plumage perfectly white, legs of an orange colour. It is well for those who think of going in for exhibition ducks to go to as many of the poultry shows as possible; then they get an idea of what they ought to be like, only of course those at the shows are fed up to such an enormous size for the purpose. In many cases they are but little use for breeding purposes after they have been shown for a season. If a good Aylesbury stock duck is bought from an exhibitor it should be purchased at the first or second show it is sent to, if not the purchaser will most probably be very disappointed in the results when it comes to the breeding season, as many of the eggs will be unfertile. It is well to refrain from buying stock ducks when they have done a great deal of winning, unless the purchaser wants them for exhibition only. An Aylesbury drake can run with either three or four ducks. Some breeders I know allow five ducks to run with the drake, and the eggs are fertile. I usually allow four, because I consider it is safer. I might mention here that those who intend showing their Aylesburies ought not to use any maize, or at any rate only a very little, for feeding, as the maize has a tendency to give their plumage a yellow cast, and also their beaks. There are many people who use maize largely, and it turns the colour of their plumage almost as yellow as the Pekins. I have mentioned this in the chapter on Aylesburies because it is so important to exhibitors. It will also be found in the chapter on feeding stock ducks.

ROUENS.

This variety of duck has made great headway in England during the last few years. No doubt this is on account of the colour of their plumage and the large size they attain to. They are especially fancied by those who are fond of wild ducks, as they are much the same colour as the latter breed. Many people in towns keep this variety of duck on account of their colour, as they do not show the dirt, and another reason is because they are so carried away by the brilliant colours of the drake. There is no domestic water fowl which comes up to Rouen drakes as regards colour. They are really a picture in themselves when they are in full plumage. The distinct colours attract the attention of every passer-by. They do not commence laying quite so early in the season as the Aylesburies and Pekins; at least, they are not considered to do so; but I have known a Rouen duck to be laying all the winter. Such cases, however, are few and far between. Of course, this makes a great difference where one wants to go in for market purposes. This class of duck is kept a great deal in the south of England. One can go for miles in some parts of the south, and you will find that every farmer has Rouens. This is specially noticeable in Kent. They do not breed them for exhibition—or, at least, only a very few—but merely for the table and eggs. I find many of the farmers prefer them to any white ducks, but of course, fancy goes a long way. I do not wonder that people, once seeing fairly good Rouens, go in for that breed. They are a beautiful shape and colour, and most people recognize them as splendid table birds. They eat a trifle stronger than the Aylesbury, more the flavour of a wild duck, and the flesh is delicious. Like most other breeds of ducks they are very hardy. Their eggs vary in colour, the same as those of other breeds. This is a most remarkable thing, because when we get certain varieties of poultry, we can depend upon the colour of the eggs; but this is not the case with ducks, as some lay almost a sky-blue egg, while others lay a dirty looking white. This is no doubt because a selection is not made of them, but I have every reason to believe that this could be altered in a few years—that is to say when a duck lays blue eggs, these should all be put aside to be sat upon, and also have a drake hatched from a blue egg from a different strain. If the eggs are selected in this way it would be easily overcome. When I say Rouens do not commence laying so early in the season, I do not mean to say I consider them bad layers. When they commence they keep on for a long period, often laying later than several of the other varieties. Their eggs, as a rule, prove very fertile. The drake should have a yellow bill with a slight greenish tinge, so that it looks like a clear yellow bronze. If it is a clear yellow it is objectionable. Some of the bills come a dark heavy colour, but these are not recognized in the show pen. The shape of the bill is shown in the illustration. It should be both wide and long. The head is of a very rich beetle-green, extending down to the neck, and should show a distinctness where it finishes off at the bottom. There should be a white ring round the neck, like a collar. This should be plainly seen. The breast is of a rich brown—what some people call a claret brown. It should be all of one colour, and not run into a grey, extending lower than the water line, so that when the drake is swimming in the pond all the breast looks of a rich claret-brown. It is only in good specimens where you get this claret-brown colour. Frequently they will become a little speckled, with lighter feathers intermixed. The lower part of the body is of a beautiful steel, what is usually called French grey.

The whole of the under part of the body should be this colour right along to the under part of the tail. The top of the back should be of a very rich greenish hue, leading right along to the tail feathers, the latter being of the same colour, with a beautiful curl. The wings should be of a greyish brown, and what are called the middle flights should have a "ribbon bar" across them, which must be of a very bright distinct blue, edged on both sides with white. The flight feathers are grey and brown. Unless they are very good specimens, white will often show in the flight feathers. The legs should be of a rich orange, and the carriage of the bird should be very noble. A really perfect specimen is a splendid picture. The bill of the duck is not quite so long as that of the drake, and is of a duller colour. Some of them are almost black. I find the colour varies a good deal, according to the season. The head is brown, with two distinct lines, one over the eye, and one under the eye. They are of a pale brown colour, and Rouen ducks should always show these two lines distinctly. The whole body is of two colours; one a very dark brown, and the other a light brown, the feathers being pencilled. They should be pencilled all over with these two shades of light and dark brown. The ducks should show dark blue in the middle of their flights, edged with white, just the same as the drake, only not quite so bright. The more evenly the ducks are pencilled the better specimens they are. As the plates of the duck and drake are not coloured they are not such a good guide to go by; but, if the readers will look carefully, they will see the outline of the feathers of the different shading. There have been a few of these, that have been well marked and large at the same time, which have fetched high prices. I have often known them fetch £10 and £15 each. The drakes during the summer months are not in full plumage and the colours are not so distinct while they are shedding them.

PEKIN.

MANY breeders who keep this variety of ducks say that they like them better than the Aylesburies. One reason for this is that they are non-sitters. Some people have an idea that the eggs from the Pekin ducks are not quite so strong in flavour as those from the Aylesburies. My opinion is that this is more fancy than reality, because the taste of a duck's egg depends a great deal on how they are fed—if possible, more so than with hens' eggs, as ducks will eat any dirty rubbish they can get hold of. They will fetch things out of dirty ponds, and they are also very fond of fish. Whenever ducks eat the latter their eggs taste very strong, so much so that one would almost think they were eating fish when they are eating ducks' eggs which have been partly produced from the refuse of the fishmonger's yard. It makes a difference if it is young fresh fish they catch alive. There are specimens of the Pekin ducks which are enormous layers, but many of the strains have been spoiled by forcing the birds for the show-pens. I knew a gentleman who had a strain of three ducks which laid 505 eggs in the twelve months—that is an average of 168 eggs from each duck. Of course this is an exception. I have kept Pekins till three-and-a-half years old, and they have laid well the last season. I do not recommend this breed to be kept over three years. I have often found they have laid more eggs the second year than the first, but they fall off a little the third year.

Of course there are duck-fanciers who go in for a particular variety. Some breeders like Pekins, and would not part with them for anything. They are certainly very showy, and are entirely different in shape from any other class. Their carriage is much like that of a goose; they stand up very erect, the breast being a long way from the ground. The latter should also have a good bow in front, more the shape of a rooster's breast; the deeper the better. They should show no keel in the front, under the breast bone, the same as the Aylesburies do. The breast should be wide, also the back, and the tail should stand very erect, just the opposite to the Aylesburies. The bill should be rather short and wide, and of a deep yellow colour. Head short and thick, just the reverse of the Aylesbury. Legs of a dark orange colour, and rather short. Their plumage is white, with a deep yellow tinge in it, similar to a pale canary right throughout. Nothing improves the colour of the plumage and beak more than feeding them on maize, and those who breed this class of ducks for exhibition should always give them a little maize four or five times a week, when they are getting their birds into condition. At the same time they should have stewed linseed given them. Their plumage is much looser than that of the Aylesbury. Pekins look very much larger than they really are. They are something like the Cochin fowls, more feathers in proportion than body. The duck and drake should have exactly the same carriage. They are not such a nervous class of duck as the Aylesbury—that is to say, if they are frightened when they are being caught, or anything runs after them, they can get out of the way much quicker than the Aylesburies do. If a stoat or rat got amongst a flock of young Pekins, it would have a difficulty in catching them, while young Aylesburies appear to lose the power of their legs, and fall down as though they were paralysed. This is a great consideration to duck-breeders where they are overrun with rats. The latter are passionately fond of ducks, and often clear several broods off in a night, and, in some cases, in the middle of the day. Where people who are fond of ducks have so many rats, I should advise them to go in for the Pekin blood. I find the young ones are quite as healthy as the young Aylesburies; in fact, they appear stronger on the legs, and fewer of them have the cramp. I do not consider the Pekins make as good table birds as the Aylesburies, as their skin has a yellow cast, which many English families object to. Pekins do not grow, as a rule, so heavy at the age as Aylesburies. I do not know why, but there are not so many Pekins which have found their way to the exhibitions these last few years as there used to be.

CAYUGA.

These are a large black duck, and the race is very little known in England. It is said by some writers that they originate from America from the Mallards. Several writers on this breed are somewhat puzzled as to where they really do originate from. Some think they were an old English duck many years ago, and were taken to America to improve upon. I will not trouble my readers by going through any pedigree of the duck, but merely show how we can make a very good Cayuga race of ducks, and improve those which we already have. If writers were to try experiments in breeding more, they would not have to wade through books and papers, and get people's opinion as to how they were made, and where this and that race of birds came from. In crossing the Rouen with the Aylesbury we occasionally get a few ducks perfectly black, not much metallic green in them, only on the wings and round the head and neck. Now, to improve this, there should be a cross between the East Indian drake and the Rouen duck. Drakes bred in this way, and mated with the cross-bred Aylesbury and Rouen ducks, will produce a fine race of Cayuga ducks, far superior to any which have ever come from America. When I was a boy I can remember seeing large black ducks; these were a cross duck between the Aylesbury and Rouen. They were to be found in various parts of the country. Because these black ducks were brought over from America and called Cayugas, they were thought a great deal of; but when they were thoroughly examined by experienced men, they had doubts in their mind as to what they really were. Many of the old judges can call to mind a very similar bird which they had seen thirty years ago. These ducks came over from America, and they said they must have been taken there and improved upon, and brought back again. They believed that they originated from our island. Writers and breeders need not trouble their brains any more as to where the Cayugas came from. Those who have bred this race of ducks, and have watched them closely, will have noticed that some of the ducks show brown pencilled feathers on their breasts similar to that of the Rouen, while at the same time in the drakes it can be seen, a few inches from the head down the neck, that the feathers are a peculiar shade, similar to that of the Rouen drake, only not so bright, but rather a cloudy heavy ring, and occasionally they will show a white ring round the neck, just like that of the Rouen. In some cases I have known them moult out almost the colour of the Rouen. In their second and third moult occasionally their plumage comes almost as white as the Aylesbury. We know black fowls, or black ducks, are liable any time after their first moult to throw white feathers. This is no new thing. Birds with black plumage, even the wild blackbirds, very often throw white wings after they are a few years old, also white round their heads, just the same as most of our black varieties of fowls and ducks. Black East Indians, for instance, will often throw a few white feathers, but not of such a distinct character as the made breeds, like the Cayugas. Those which were imported from America are exactly the same shape as the cross I have mentioned, viz., the Rouen-Aylesbury. Even the writers on this breed say that they are in shape similar to the Aylesbury and the Rouen in their pure state. I do not mean merely to say that it is possible the Cayugas can be made up with the crosses I have mentioned, but that I am sure they can. I have produced many such birds myself, and have seen others do the same. There have been two pens of black ducks just round my neighbourhood for the last five years of the same cross, produced by my own birds. People call these ducks Cayugas. Certainly they are good layers; they naturally would be, because they are made of the right stuff. Those who have Cayugas have a difficulty in keeping the bright plumage, as it often gets dull; but if they will try the cross I recommend in the commencement of this chapter with a large Rouen duck and the largest East Indian drake that can be procured, their pens of Cayuga ducks will be wonderfully improved. I do not make this statement to run this variety of duck down, rather the other way about. They are really a splendid duck, good layers, and produce fine table birds, and as the colour does not show the dirt, they can be kept in towns, or in very small places where they are liable to get dirty. It is a breed which I have not kept very many of, but I hope to keep them more than I have done. The reason I have not kept as many as I might have done, is because they do not breed true to colour. When a person sells ducks' eggs, and represents the stock to be all perfectly black, and some of them come out with white feathers occasionally, and sometimes some of them come with a few brown feathers, it makes people think at once that those who have sold them had an end in view to deceive their customers. I mention this so that any purchaser who may be induced to try this breed must not be disappointed when occasionally they do not breed perfectly black ducks.

Cayuga ducks should be black all over, the wing bars a purple green as well as the head. There should be a good deal of colour on the head and neck of this purple green. Many of them come with a dark, heavy plumage, a kind of sooty black. For exhibition purposes they should have a beautiful metallic green cast all over their plumage, the brighter the better. The bill should be of a dark colour, almost black. Legs of a very dark orange colour, rather inclined to be of a sooty shade, not bright, like those of the other ducks.

MUSCOVY.

THIS class of ducks is quite distinct from any other; in fact, so much so that they object to other breeds very much when they are allowed to swim in the same pond. They very seldom take any notice of any other variety of ducks as regards the intermixing of breeds. We seldom come across this breed of ducks, except in print. They appear far less taken up as a domestic duck than any other variety. There is a little piece in "The illustrated book of poultry" by Lewis Wright on this breed. The representation he gives as regards their character is rather misleading, as he says that the drakes will persecute everything they can get hold of. I have bred hundreds of these ducks and had a good many drakes for stock purposes, and it is only occasionally I have found one which will turn against the other varieties of poultry. A Muscovy drake has double the strength of any other drake. When they are allowed to run with fowls they will occasionally fall out with a cock bird, and have a sharp fight, just the same as any other drake will do; but I have never found them dangerous in any way, although they used to run about with my chickens; in fact, I have had more experience with the Muscovy drakes than I have with the ducks of that breed, as I have used them very largely for crossing with the Aylesbury. It must be understood that they do not care about crossing with other ducks if they are allowed to run with ducks of their own class at the same time. They should always be taken away from the ducks of their own breed before the breeding season, or the result will be a very poor one if cross-bred ducks are required. When I have been visiting various parts of the country I have noticed that cottagers sometimes keep a duck or two of this variety for pets, as they are very tame and docile.

Then, again, Lewis Wright speaks of the Muscovy duck being loose in feather and looking miserable. I have never seen one loose in the feathers yet, except when they have been moulting. Mr. Wright also calls them bad layers. This is altogether out of place. I have not met with a bad layer yet, although no doubt there may be a few, but taking them on the whole, they will lay quite as well as our ordinary farmhouse ducks do, or at any rate as well as they used to. Certainly, people have improved their breeds in the farmyards, even in the last few years I have noticed this, but the Muscovy duck will lay more than the ordinary farmhouse ducks used to thirty-five years ago. This breed of duck are rank sitters, they do not lay many eggs before they come on broody,—in most cases from eleven to seventeen eggs; but I have known them lay thirty eggs without becoming broody. I have heard of them laying two shelled eggs in a day, and can remember one Muscovy duck which was known to lay thirty-three eggs without missing a single day. While we get such specimens as these I do not see how we can call them poor layers. No doubt inferior layers are to be found among them, the same as there are among all other varieties, especially as they have never been bred for this purpose; but with care and selection I feel sure they can be improved upon in this respect, and become a very valuable variety of ducks to keep. They will often lay right up to October, I have had them bring young ones off in November. Their appearance is quite different to other breeds—they look quite a curiosity. They please many people, especially gentlemen who have large grounds. Two or three of my friends who keep a flock of them say they fly round just the same as pigeons do, high up in the air, but always take care to come back to their proper home. My ducks will fly round just the same as wild ones will. There is one thing—when you go to catch them the strength of their wings (especially in the drakes) is very great, so much so that it would almost break one's finger if the butt of the wing were to catch it. Drakes do not like to be handled, although they are so tame; they do not object to be stroked, but when it comes to being handled they will fight for their life. There is a peculiar smell about the drake, similar to musk, while alive, which some say the variety take their name from; but when they are dead this musk smell is not noticed. The flesh of them is rather dark, similar to that of the wild duck, and the flavour of them is very much the same. Perhaps it may be a little strong, but some people are very fond of it.

The Muscovies are very hardy as young ones, and grow quickly. The carriage of both the duck and drake is very peculiar. They have a long flat body, very short neck, and long tail, standing straight out; they have also very long wings. They vary in colour; some are pure white, but these are very rare. The drakes usually come black and white; but the black is really a beautiful metallic green; the white is also very bright. The colour is quite distinct, and not intermixed, a large patch of white and a large patch of black. If anything puts them out, or they are a little frightened, they have the power of raising the feathers upon their heads. Their bills are the most peculiar feature, especially in the drake, the latter have a clear scarlet place round the eyes free from feathers. The base of the bill has a band of scarlet round it. The head of the drake is very large. The duck shows the same head points as the drake, only they are much smaller, and the red round the eyes and face is smaller than in the drake. The bill of the duck is darker in colour, and not so brilliant as that of the drake. One extraordinary thing about these ducks is that their eggs take a week longer than those of any other variety to hatch, viz., thirty-five days. Sometimes they may be out a day, or even two, before the time, just the same as those of any other breed of ducks. It depends a great deal upon the heat the eggs are kept at during the period of incubation, and also the freshness of the eggs.

It will be noticed in this variety that every step or two they take they keep thrusting their heads forward; this will be observed more particularly, perhaps, in the drake. Some of the ducks come black and white; but in some cases I have known all the ducks to breed perfectly brown, almost the colour of the Rouen. The drake is very much larger than the duck; indeed, there is so much difference between the size of them that an inexperienced person, who had not seen them before, would scarcely believe that the duck belonged to the drake. I have had two drakes turn the scale at 12 lbs. each after they were two years old. I find these drakes really grow up to about the third year. Of course, they do not all come up to this weight. The weight is very good if they weigh from 7½ to 9 lbs. the first year. I do not care about getting the drakes too large, as they become too heavy for the ducks when they grow to such an enormous size. The one illustrated here, in this little work, turned the scale at 12 lbs., and the duck at 7 lbs. I believe this variety of duck will become more popular in a few years. I shall try to encourage people to keep them, as well as try to improve them as layers. Some of my readers may say, "How can they be improved upon when they lay every day between the periods of sitting?" They can be induced to lay earlier in the season, and keep on laying later in the summer; that is where the improvement can be made. They do not commence laying as a rule till about the 1st April.

BLACK EAST INDIAN.

In this little work on ducks I do not attempt to go into fancy varieties, only breeds for practical use, and also for exhibition, for the benefit of those who prefer them. I may therefore be going a little wide of the mark when I give a chapter on the Black East Indian Duck. These are kept very much in England, for their beauty, and at the same time they are good birds for the table, though they are so small. They are what I call a half-way duck, kept for fancy as well as for eating purposes. As a rule people who keep fancy ducks never think of eating them, but it is not the case with those who go in for the East Indian. There are some high-class people, where there are large families, who seldom eat any other variety of duck except this, as they keep them in large flocks. They are a very sweet-eating duck and most delicious in flavour, but of course they are remarkably small, and not much good to put before a hungry family. They are a very choice dish, and thought much of by those who are epicures. They are usually eaten as a sweet morsel and not put on the table to make a hearty meal of. They are fairly hardy. They will bring up their own young and require no pampering. Even when they are exhibited there is no need to prepare them for the show pen, as they do not require to be got up to any extraordinary size. In fact, it is really the other way about. They are much like the Bantam fowl. The smaller they are (so long as all their other points are correct) the more valuable the birds are, the greater the beauty, and the higher they stand in the judge's opinion. Their plumage is of a beetle green all over, having a very bright metallic cast. Their bills should be dark. Opinions vary as to the shade they should be, but I like to see them a bronze green of a very dark shade when they are young. The legs and feet should be of a dark colour, showing a little rough, as there is a kind of pinkness under the skin. After they have turned eighteen months old, the pink begins to show under the skin more brightly. The legs of very old ducks become almost of a dark orange colour. When this breed of ducks are required for exhibition, if a little stewed linseed is given to them mixed in some meal about ten days before they are exhibited, it puts an extra gloss upon their plumage, and will well repay the owner for the extra trouble.

CROSS-BRED DUCKS.

Crossing for table a profitable thing—The old and the new crosses: comparative advantages—Indian Runner crosses for laying—Good feeding for good egg results—Pekin-Rouen—Muscovy-Aylesbury—Wild-Rouen.

ROUEN—AYLESBURY.

THERE are many people say that it is very absurd to cross ducks, as it spoils the breed, and not only so, but it lowers the value of them if they are required to sell. People put it down as a waste of time and money, besides spoiling the ducks. Now it is quite true that the offspring will not make such long prices as pure-bred specimens, but we must bear in mind there are always two sides to every question, and I will try to deal with both in this chapter. Some breeders I know would not think of crossing their ducks, or keeping a cross-bred duck; but remember there are twenty people who rear ducks for their own consumption and market purposes only, where there is not more than one who merely keeps good-bred ducks for exhibition. I do not think I exaggerate when I say that there are one hundred ducks reared for the market and for one's own table, to every one duck that is brought up for showing purposes. It is well, therefore, to see which are the best for all-round purposes. We will suppose a person only wants to keep, say a drake and three ducks, or four, just to get the eggs from for hatching for table. In the first place, a pen of cross-bred ducks are cheaper to buy, and lay more eggs than many of the pure varieties, and the young ducklings grow just as fast, and, as a rule, are quite as heavy as those from the pure strains when they are ready to kill. Some of the cross breeds will run $\frac{1}{2}$ lb. heavier; I have known them to weigh $1\frac{1}{2}$ lbs. heavier than the pure breeds, when the latter and the cross breeds have both been running together. I have tried this experiment for several years. Perhaps I had better explain what I mean, so that my readers may see the advantage of cross-bred ducks in a back yard, and where they have not much room or accommodation. In former years the best cross we knew of was the Rouen-Aylesbury, as they grew to a good size, and were fair layers. Indian Runners, however, have beaten every other breed in their pure state as regards their laying qualities, but they are somewhat small ducks. There has always been a scarcity of eggs to produce early ducklings, but I introduced Indian Runners into the South of England, and through the Midland counties, where they were not known a few years ago, and have recommended them for crossing with the Aylesburies or any other large ducks, no matter whether they be cross bred or pure. They improve the laying qualities very much. The best plan is to have an

Indian Runner drake with four ducks. My experience is they do better with the Aylesburies, but they can be used with other breeds. First put an Indian Runner drake with Aylesbury ducks, which brings most of the young ones brown and white, save the ducks, and the following year put an Aylesbury drake with them; that will bring size, and as regards their laying qualities, twenty of them will lay more eggs than 100 pure Aylesburies during the winter. Not only so, but the young ducklings will grow faster than the latter, and at nine weeks old will usually run about the same weight. There will also be less deaths amongst them. I am pleased to say where people have tried this experiment in the duck counties, it has answered splendidly. The Rouen-Aylesbury is another good cross for laying purposes, and for producing good table birds. To get a first cross it is necessary to have a drake of one breed and ducks of another, say, for instance, Aylesbury ducks and a Rouen drake. If good framed Aylesbury ducks are selected, and the Rouen drake is mated with them, their offspring will be very hardy, and less liable to be struck down with the hot sun. The young ones will also begin to lay earlier, as a rule, and keep on much later in the season than pure bred. Cross-bred ducks do not shed their feathers so early, by about a month or sometimes more, that is to say, if one has a pen of cross-bred ducks for breeding from, as well as a pen of pure breeds, taking the whole of the season through, the cross breeds will lay as a rule from twenty to thirty eggs more per duck in the twelve months—that is, if they are both bred from the same stock ducks, only, of course, crossed with another drake. As I have said, they commence laying earlier in the season, and also keep on later. No doubt this is because entirely fresh blood is introduced. The same thing occurs when the Aylesburies and Pekins are crossed. It usually strengthens the laying organs, and they will lay far more eggs during the cold weather, early in the season, just at the time when eggs are required. Those who have only a little money to spare, and wish to keep a small pen of stock ducks, should try and pick up a few first-cross Indian Runner-Aylesbury or Rouen-Aylesbury. I have tried these two crosses during the last few years, and the Indian Runner-Aylesburies have laid from 30 to 35 more eggs per bird in a year than the pure Aylesburies.

I am aware that it pays me far better to sell a pure-bred duck than it does to sell a cross-bred, but that is not the thing. I find out which are the best to keep, and then give the public the benefit of my experience.

To those who only have a small space to keep a few stock ducks in, and who, perhaps, have no water, or at any rate very little for them to swim in, I would say buy brown ducks, they do not show the dirt so much as the white. At the same time, where there is a pond or brook for them to go to, white ducks will usually keep themselves clean. During the last few years I have kept white ducks on grass where they have had no water to swim in, nor even sufficient to wash themselves in. They have kept quite clean, and the eggs have been as fertile as those from ducks which had a pond or stream to go in, and they have laid quite as many of them as the latter. Usually I do not recommend people to keep stock ducks from the water if they have plenty for them to go in, but those poultry keepers who are not fortunate enough to have a pond or stream can keep and even breed from them successfully without. I have had ordinary stock ducks of various breeds lay from 120 to 150 eggs per year with no water to run in, but this is rather an exception, and not the rule, with ordinary ducks.

We do not complain if a duck lays 100 eggs in twelve months. When I first kept ducks, over 30 years ago, they did not average 45 eggs each in the year, and more often not more than thirty. In the old fashioned way of keeping them it was said that it did not matter what the ducks were fed upon, as they would eat any dirty rubbish. It must be remembered that an egg contains a certain amount of nutriment, and, unless the duck first partakes of something which is nutritious, it is against Nature to expect a number of eggs. The same thing applies all round. If a person has a cow which gives a quantity of good rich milk, and it is fed upon mouldy hay or inferior food, the supply of rich milk will very soon begin to decrease. Again, a person may have a piece of land; but, if he does not cultivate it properly, and put sufficient manure upon it, the returns will be very small, and it would be partly a waste of time. It is just the same in keeping any kind of stock. If they will not pay for being kept well, they certainly will not pay for being kept badly.

So it is with the ducks; the better they are kept and managed the greater the harvest. I have tried all crosses of ducks, but have found those I have mentioned to be the most profitable, when everything is considered, because they can be bred from again so much better, and they make nice-looking ducks.

I should mention that in breeding the Rouen-Aylesburies some of them come pure white, just the same as a pure-bred bird, and it is almost impossible for one to detect that they are a cross at all. It is just the same when they are crossed again; that is to say, that when a pure Aylesbury drake is mated with the Rouen-Aylesbury ducks, many of the ducklings come white, a few black and white, and occasionally one may appear brown, but only a few. I like the look of the first cross much better than the second cross, though the ducks themselves are no better in quality; the second cross are quite as good for laying and breeding as the first cross, only the former are not quite so uniform in colour. When the second cross are bred from again, I recommend a Muscovy drake, then they produce very large ducklings. Or, if it is more convenient, a pure Pekin drake can be used; but in all cases see that the drake is pure, as the results are so much better. A good-sized drake should always be selected.

To those who intend breeding ducks for sale for stock purposes, I should recommend pure ducks of whatever breed they may fancy, as in that case pure-bred ducks would pay them best. Then again, some people prefer seeing pure ducks on their premises; these, by all means, should keep them.

PEKIN—ROUEN.

THERE are some people who keep Rouens, and are rather disappointed that they do not lay earlier in the season. If such people do not object to cross-bred ducks, they should run a Pekin drake with them, as it will increase the laying very materially, as this cross will often lay a month or six weeks earlier in the season than the pure Rouens. If a good number of this cross are hatched, many of them will come all brown, just the same as the Rouen, only of a trifle lighter shade. These can be selected for breeding purposes for the coming year. To breed from them again an Indian runner drake should be put with them. This will bring very fine table ducks, and they will grow rapidly, and make excellent layers. I have had some of the Rouen-Pekin ducks lay six eggs a week each for months. As many of the readers of this little work will know, eggs which are laid early in the season are very valuable if they come from large stock. No matter whether they are pure or cross, it is the eggs that the duck people want if they are to make a good price for the young ducklings early in the spring. I write on the crossing and re-crossing for the convenience of those who cannot afford to pay for high-class pure stock ducks, or those who may have the stock by them. I do not recommend anyone to pick up small cross-bred ducks, very often crossed with the wild or the Irish ducks. The latter breeds are mixed a good deal with the wild, and it is a disappointment to the purchasers when they expect a fine large duckling at eight or ten weeks old, to get miserable things weighing from 2¾ lbs. to 3 lbs. each. It is entirely a waste of time. Therefore every breeder, or those who buy a few eggs for the purpose of producing ducklings for their own table use, or to sell, should take care to get their birds from good large stock ducks.

PEKIN—AYLESBURY.

SOME readers may ask, "Why do you give a chapter on the Pekin-Aylesburies, when you recommend Rouen-Aylesburies in preference for useful purposes?" For this simple reason—many readers might be very disappointed were I not to do so. It frequently happens that persons have a few Aylesbury ducks and a Pekin drake by them. This, of course, saves expense in getting others. Those people who have already got their stock by them, if I were not to mention how they come, and what they are like, might think they were not worth breeding from. I always endeavour to save people expense rather than run them into it. I say, therefore, to those who may have such birds by them, mate them together, as they make wonderfully good layers, quite as good as the Rouen-Aylesbury, if not better. At the same time, I do not like the cross so well, because they do not come out quite so distinct as the Rouen-Aylesbury; neither do I like them so well as a table bird. They look like poor-bred Pekins, or poor-bred Aylesburies—nothing of a distinction about them; that is why I do not like them. They are very profitable for laying purposes, and make good table birds. A few of them have a yellow cast over their skin; but when they are killed, you could not tell the difference between many of them and pure Aylesbury. Some of this cross are often sold for the pure Aylesbury when they come out with flesh-coloured beaks. This is another reason why I do not recommend this cross. At the same time, those who have the stock in hand should by all means use them.

MUSCOVY—AYLESBURY.

This is a cross which I strongly recommend, especially to those who have had no experience in duck-rearing. When the Muscovy drake is used with the Aylesbury ducks the results are splendid, both for laying and table purposes. Some say the flesh of the pure Muscovy is not good, as it is a dark colour; but the Muscovy-Aylesburies are splendid table ducks. I have had this cross for years for my own consumption, and I have got them to a greater weight at ten weeks old than any other variety I have ever bred. They were more like geese than ducks, and the flesh is most delicious. I bred a number of them in August one year, and the smallest of them weighed about $7\frac{1}{2}$ lbs. A few of my customers begged me to sell them for the table. They offered me 5s. each for them, and, as I thought that was a little more than they were worth for my own consumption, I sold a few to different people who visited my poultry yards. I sent them to various parts of the country. It has been said that if the Muscovies are crossed with other ducks the progeny will not breed again; but this is wrong. As a rule, they breed speedily, and make good layers. The eggs from the Aylesbury ducks, crossed with a Muscovy drake, do not turn out quite so fertile, as a rule, as the eggs from other crosses of ducks; but the result in the ducklings, although not so many in number, more than satisfies those people who try them. Many of them come brown in colour, and occasionally a few black and white, and now and again a white one crops up. They are quite a different shape from any other duck. This is because the Muscovy is so different in shape to any other breed. The Muscovy drake can be put with either Pekins or Aylesbury ducks, but I prefer the latter, as the flesh is so much better when crossed with the Aylesbury. Some of this cross have been sent to the London markets and realised very high prices, as they are much larger than the Aylesbury ducks. The eggs of this cross, when a Muscovy drake is used, will go from twenty-nine to thirty-one days, instead of twenty-eight. The eggs of a pure Muscovy are five weeks in incubation, while those of other ducks are only four. Of course they may hatch out a day, or even two, before the time, just the same as any other variety will. The Aylesbury ducklings will sometimes hatch in twenty-six or twenty-seven days, but the proper time is twenty-eight days.

There are very few Muscovy ducks in England, therefore it is rather difficult to obtain drakes for crossing purposes. When it is convenient, the Muscovy ducks can be crossed with the Aylesbury drakes, in which case the eggs may go even a day or two longer. Often they will not come out before the thirty-second day. The result will be more satisfactory when the Muscovy drake is used, as the Muscovy ducks are very particular, and do not care to mate with any other race.

WILD—ROUEN.

There are many people who are very fond of the taste of Wild duck, and prefer it to any other. The flavour is different to that of other breeds, with the exception of the Rouen. Many gentlemen in England who breed a large number of ducks for their own consumption, and to give away among their friends, have nothing but the Wild and Rouen cross. To commence this cross, the best way is to get small Rouen ducks; they must not be large, or the eggs will not be fertile, as the Wild drakes are very small. If nice small Rouen ducks are selected for the purpose, and from four to eight ducks turned into a large pond with from two or four Wild drakes—there should always be about two Wild drakes to four or five ducks—the offspring come exactly the colour of the Wild, as the Rouen are almost the same colour, and every duck and drake appears to be marked perfectly even. When gentlemen once get this cross they breed again, with the Wild and Rouen drake with ducks of the same cross. I know a gentleman who has bred them in this way for many years. The ducks and drakes run from 3 to $4\frac{1}{2}$ lbs. each, and the flavour of the flesh is very good indeed. To get fresh blood it is necessary to obtain drakes of the first cross from some other strain crossed in the same way. These ducks are very hardy, and do not lay, as a rule, till the warm weather comes. They will make their nests, lay their eggs, hatch and bring off their young, without any difficulty; they usually hatch two broods a year each. They want no attention at all, except the feeding. They look very handsome where one has a large piece of water, and also come in very useful for table purposes. This cross can fly like the wild ducks, but, as a rule, they do not go far away from home. Occasionally they will fly a few miles, if they are not pinioned, but they usually come back again to their own pond. This is not a cross I would recommend to anyone who brings up ducks in confined runs, only to those who have large premises and extensive ponds or streams. In such cases they will be delighted with this cross. Of course, where persons only want to bring up broods for their own consumption, they can be brought up in confined runs; but they are not so profitable as other breeds.

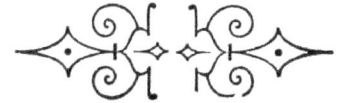

THE HISTORY, SPREAD & DEVELOPMENT OF INDIAN RUNNERS.

Indian Runners—Laying points of Indian Runners—Indian Runners for crossing.

A FEW years ago I heard of a duck called "Indian Runner," and had a great deal of correspondence respecting the breed with people in the North. When they described to me what the birds were like, I said at once they were cross-bred, but sometime afterwards I was travelling through the North, and came across several pens of these wonderful ducks, which caused me to alter my opinion. They were reported to be wonderful layers, and directly I saw them I could see they were pure, and a distinct breed. I at once purchased some, and went right through Cumberland and Northumberland to learn all I could about them, making a few purchases in various districts, at the same time, so as to have birds unrelated. I found the ducks were small, and there were two distinct colours in them, though they were all alike in shape, and had the same carriage. Some were white, intermixed with pencilled brown, and the others were white also, but instead of being of a pencilled brown (similar to that of a Rouen) they were more of a fawn colour, or very light brown, almost bordering upon buff. I have tried both colours, and found them extraordinary layers. As I found no difference between them in that respect and they had the same shape and carriage, I came to the conclusion both were pure.

I have traced these ducks back in Cumberland for over 60 years. There I met with a lady, over 60 years of age, who told me her father kept them as long as she could remember, that was in 1890, and her father said he had kept them for many years, so they may have been in existence as many as 80 years or more, but I am sure of over 60. I never heard of them, or saw any in the Midland Counties or South of England until I introduced them there from the North. They can now be found in almost every village and town in the country. Those who have tried them say they have never known any other ducks to lay as many eggs as Indian Runners.

Many of these ducks in varions parts of the country have laid over 200 eggs each in the year, some of them have kept on all through the Winter. Several people have told me they have laid more eggs than the fowls. I have had them lay at five months old, and the old stock ducks have often laid an egg each day for from 20 to 40 days right off without missing once. One laid 75 eggs, and another 86 without missing a single day. These are, of course, exceptional cases. I knew a poor woman who had 13 of these ducks, and she said she had an average of 12 eggs per day for four months. I was very much interested in the poor old lady's ducks, as she was almost blind. She showed me bacon hanging up in the house, and said she bought two pigs with the profit from the ducks, fattened them, and sold the hams to buy boots for the children, and had four sides of bacon left for the use of her own family.

A gentleman in the North had 18 of these ducks, which had shed their feathers during the summer, and from the 14th August up to the 14th November they laid over 1200 eggs; that was in the three months which, as a rule, ordinary ducks do not lay any eggs in. I do not remember being a month without eggs at any time of the year from my Indian Runners. The drake belonging to the pencilled brown ducks is of a slate colour, and of just the same shape and character. He should have a white neck, and slate coloured head and breast. The top of the wings should also be that colour. The more evenly marked the birds are the better. Good specimens appear to have a white band round their chests, so distinct are the markings; for ordinary purposes it does not matter if they are not so evenly marked. The ducks are the same colour as the drakes. When they are young they have an orange yellow beak, but as they get older it becomes darker, particularly in the ducks. The other variety are quite different in colour, though they are Indian Runners. These birds are splendid foragers, and some parts of the year will get the whole of their living when allowed a good range, they are very fond of slugs, worms, frogs, &c,

They will also do well in confined places. I have known many people who have kept them in small back yards, and they have laid during over ten months out of the twelve, but it would take up too much space to give the number of eggs and the various reports of those who have tested these birds in this little book. Both ducks and drakes should be tightly feathered, very erect in their carriage, and have a long neck.

They get about in a different fashion to other ducks, instead of waddling they run straight off. No doubt that is one reason why they are called Indian Runners. They are always on the move. In good specimens the head should be fine and very flat, more so than that of any other duck. The eyes are very near the top of the head.

The beak should be strong and fairly broad, coming straight down from the skull. Some people call the head a wedge head. Perhaps the illustration of the White Indian Runners will give a better idea of the shape of the head than that of the coloured, as my artist has drawn it more correctly.

Many cross-bred Indian Runner ducks have been sold as pure, because the drake, when used for crossing, stamps his image upon the progeny so plainly. I have come across many people who have told me they possessed Indian Runners, but they have turned out to be only half bred. If a person once sees these ducks he never forgets their beautiful carriage and graceful movements.

Indian Runner drakes are the best breeds to use for crossing purposes to improve the laying qualities. Not so much for size, because in some cases the progeny do no grow quite so large. As a rule, when the Indian Runner is crossed with Aylesburies or Pekins, the half-bred ducks do not vary a half pound in weight from the two latter breeds in their pure state: we have known them grow even heavier. To some people this would make 35 per cent. difference above what their ordinary ducks have ever made. It seems a mysterious fact that such valuable ducks have been in our country for over half a century and have not spread before. No doubt they would not have done so now, had it not been for poultry literature.

Although Indian Runners are small, it pays a person to keep them only for their eggs. They are wonderfully plump also. When they were first introduced in the Midland Counties and the South of England the poulterers complained of them being so small, but they find they are of such fine flavour, very much like that of the wild duck. There is a great demand for them, and there is almost as much meat upon them as a big framed duck, but they are not so fat. India Runner ducklings are very hardy, and no trouble at all to rear.

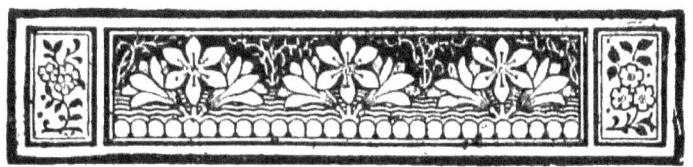

WHITE INDIAN RUNNERS.

New breeds introduced—Reception of new breeds by captious and badly-informed people—Points of White Indian Runners—Laying qualities and characteristics of Indian Runners generally—Beware of fraudulent imitations—Activity of Indian Runners.

DURING the last few years many of my readers will remember I have brought out six different varieties of poultry, but they may be somewhat surprised to find a new variety of ducks introduced. Some people take to a new breed or variety of fowls at once, particularly if the birds have good laying and table qualities. They think it is nice to have something fresh.

Others say at once, "Why do we want so many varieties?" and they write against them at once, particularly if they have been to a show, and seen new varieties there.

Yet, strange to say, when people see new varieties of flowers at exhibitions they hail them with delight and say, "how lovely."

I have now brought out a new variety of ducks called "White Indian Runners," and no doubt some will take them up with pleasure, whilst others will try to condemn them. I have several reasons for introducing these ducks. When Indian Runners were first introduced into the South of England, and the Midland Counties, many of those who took them up had them principally for their wonderful laying qualities, but said they did not like the colour as they were a mixture of brown and white. They remarked what pretty ducks they would be if they were a self colour. I at once set to work to breed a white variety, and have succeeded in producing White Indian Runners without a drop of any other blood in them. They have simply been bred by careful selection and mating. I bred an immense number of them this past season, but unfortunately have had a very large number taken away by rats, which has reduced my stock considerably. Probably the first question my readers will ask is, "Do they breed true?" Not quite, but they are all Indian Runners. Some come perfectly white, and a few come spotted, and will do so yet for two or three years.

It must be remembered, however, that when any new breed or variety of fowl is brought out, for which there is a good demand, fraud is practised directly, and I quite expect people will buy up any white ducks they can get hold of and call them Indian Runners, although in many instances they will be altogether different to the real breed. The correct points should be wedge shaped head and beak, the latter of an orange yellow, and legs of the same colour. The eyes, as in the coloured variety, are set higher in the head than any other breed of ducks. They are small ducks, very compact and close in feather, neck should be thin and long, and they have the same carriage and shape as the coloured variety. In good specimens the plumage is perfectly white, without a shade of yellow, if there is a yellow cast that shows Pekin blood. Refer to illustration, which gives a fair idea of what these ducks should be like. They lay equally well as the coloured birds. Some of my friends who have reared many of them for me say they lay better, but I do not think that possible. Perhaps if a person were to take forty white and forty coloured the former might lay more eggs, because I have taken great care in selecting my breeding stock from first class layers, but I say do not expect more eggs from the white than the coloured ones, if they lay more so much the better.

I am quite confident this new variety of duck will give great satisfaction, and if a few coloured ones do crop up they are all Indian Runners.

Up till the time I introduced this new variety, so far as I know, no other fresh breed of ducks had been brought out in England except Aylesburies, and it is 80 or 100 years ago since they were introduced. Those who go in for Indian Runners (whether coloured or white), will have the best breed for laying they can have, and if there is space for them to roam about in they will pick up half their food, particularly during the spring and summer. They are wonderfully active, so much so, that it is very seldom a fox can catch them, as they will fly to their home like a wild duck. If they stray away from home they always find their way back. They are quite different to Aylesburies and Rouens in this respect. Both these breeds, particularly the former, if they see a rat or fox, appear to lose the use of their limbs, and fall down so that an enemy can catch them, but Indian Runners are so nimble that unless they are very small they are up and away at once. They are rightly named. The eggs of Indian Runners are very fertile.

FATTENING.

Fattening an important part of the duck industry—Food that will fatten—How to treat ducks when fattening—Killing: how to select for—Shelter for ducks.

FATTENING ducks is a very important item with those who supply the markets. These are not fattened in the same way as fowls; ducks are not required to be crammed or fed by a machine the same as the former. The ducks will usually cram themselves if they can only get at the food. It really depends upon how the ducks have got on in the way of fledging, and the time of the year, as to exactly when to commence fattening them. Young ducklings should always be fed well on nutritious food, so that they are in good condition; but to plump them up for the table, to get them in tip-top condition, they should have extra feeding the last fourteen days of their existence. The food given them may vary, but should in every case be of a fattening nature. Rough fat should be put in their food; this may be bought at the butcher's. Very often when it is bought there is a little lean on it; this should be boiled down or cut up very fine,

It is well where there are a large number of ducks kept to run it through an old sausage machine, it is the quickest way; then, if boiling water is poured upon it, it saves the trouble of boiling; then mix it up with the meal, and add a little salt with it. When the rough fat cannot be procured, granulated meat should be used; it is also a good thing to use the fattening powder about four or five times during the fattening period. This sharpens their appetite, and helps them to digest their food, and is really very little trouble, as it only wants mixing with the other food. Good barley meal and ground oats mixed is one way of feeding. I generally use a specially prepared duck meal of my own, which contains all meals of a fattening nature. I find this is excellent with a few other little things added. Where a farmer has peas which he can grind, or is able to get peameal at a fairly reasonable price, that can be mixed with barley meal and sharps. This is of a very fattening nature. There are many who rear and fatten young ducklings who would not go to the expense of buying biscuit meal, but if only just a little is mixed with other meals, it answers so much better, particularly for early ducklings, when they make a long figure in the market, as they grow so quickly when it is used. There is no stickiness about it, and it well repays the owner for the expense.

When the young ducks are put up to fatten they should not be allowed to run about, except just in a yard or shed to feed, and should be kept very quiet. If there is a window in the house where they are shut up, and it is sunny weather, it is well to put a curtain just to shade the light from them.

If it is a little dark they usually rest better. Some people only feed them twice a day, but I like to feed mine three times; then they do not gorge themselves quite so full. Never allow any water to stand by them, but let them have as much as they can drink when they are fed. Where small potatoes can be bought cheap, boiled up and mashed fine, and mixed with the meal, it forms a good food. When the weather is cold the food should always be given to them warm. I really like to give warm food to mine even in the summer. When warm food is given them early in the spring they grow almost half as fast again and fledge quicker. In all cases I like to feed them in troughs. If a flock of ducklings run together, some of them are sure to be a little more forward than the others.

It is well to begin to fatten them when they are from seven to eight weeks old. Sometimes when they have done extra well they are ready for the table when they are between eight and nine weeks old. In such cases, when they are so forward, of course they want fattening earlier. The larger ones can be killed first. They put on a great deal the last fifteen or eighteen days of their existence. When ducklings have not got on well as regards size, they are not ready for killing before they are about ten or eleven weeks old. In such cases they should not be put up to fatten till they are eight weeks old. Persons must be ruled, to a certain extent, by their own judgment in this. Those who have fattened two or three lots will soon know exactly when to begin. Always give them a little fine grit in their water about twice or three times a week.

When a person has a brood of ducklings, and kills them off one or two at a time for his own table, he should never leave the last over twelve weeks old, but should have them all killed before they are quite that age, if possible. When they are let go over that age they begin to lose weight, as they commence shedding their feathers. If people cannot eat them all themselves quickly enough, they should sell those they have over. In hot weather it is not necessary to shut ducklings up to fatten, they will do equally as well under shady trees if a piece of wire netting is run round so that they cannot get too far away. When ducklings are allowed to run on grass the skin is never quite so white, but equally as good in flavour as when they are shut up. It is difficult to provide sheds when one has several hundred ducklings. If there are no trees to act as a shade, it is well to put some old canvas on frames to shelter them from the sun in the middle of the day.

GRIT.

Sharp Grit for Ducks: let it be sharp—Flint Grit *versus* Sand, Shell, and Shingle—Aylesbury Ducks: Fallacies and Facts—Dangers arising from insufficient sharp grit.

THIS may appear a very peculiar thing to give ducks, as there are many people who keep them and never think of supplying them with the material to masticate and digest their food. But it is really as necessary as teeth are to the human being, or any animal. The feathered tribes which eat corn have no power of masticating their food except in the gizzard, and that cannot be done properly unless they are supplied with the proper material, which must be of a hard, rough, jagged, and sharp nature. It cannot be too hard nor too sharp for this purpose. I always consider that grit means health and happiness to the feathered tribes. This is where so many people who keep ducks and fowls make a mistake, and failure is the result instead of success. It is really impossible for them to thrive unless they are supplied with sharp grit. The gizzard of a duck is so wonderful that they will often swallow a piece of a broken nail, or any very hard substance, if they can get hold of it, when they are kept short of sharp grit. I have found the edges of a nail worn perfectly smooth, and as bright as a piece of polished steel. Ducks will often swallow pieces of glass, and when they swallow them the edges are as sharp as a knife, but when they pass out of the duck, as a rule, they are as smooth as a piece of polished marble; every particle of the edge is worn off, just the same as it is in the fowl's gizzard. Anything sharp, such as flint stones, gets worn down quite smooth. This shows what power the gizzard has of masticating the food when hard material is inside it. Of course it is the friction caused by the hard material rubbing together, and the action of the gastric juice, which makes it smooth. Now, I am often asked, "What is the best grit to give?" I have every reason to believe flint grit is the best; but it is rather difficult to get in some neighbourhoods, and consequently there are substitutes which will do. Some people use granite, others a quantity of broken glass; either of them is a very good substance as long as they are not too large. The flint is much sharper, and does its work better. Grit for ducks should not be so large as that which fowls have. I always prepare grit especially for ducks separately from that for fowls, and never use anything but the broken flint. Some people bake the flints so that they break more easily. This is simply spoiling a good material, as it is so much softer, and does not answer nearly so well. The edges, instead of being hard, are soft, and as soon as they get in the gizzard they are ground down to powder at once. Those who keep ducks, and are short of grit, should always break up all their white glass and old china, or anything of a sharp and hard nature. Ducks and fowls apparently exist for the purpose of turning everything to good account. There is a great deal of refuse which would be otherwise wasted from almost every house; whether by the rich or poor, it would be wasted were it not for the feathered tribes. Young ducklings should always have grit given them from the time they are two days old. Road scrapings, so long as there is any hard and sharp substance amongst it, is a capital thing for ducks. In the gutters along the country roads, after there has been a heavy storm, there will be found a great deal of sharp grit, that which has been broken up by the cart wheels. If swept up and put in the ducks' runs or water, this will be found very useful to them. I prepare a fine grit especially for young ducklings. Some people use sea shingle, or sea sand. It is only a waste of time to give ducks or poultry this material, as the sharp edges are worn off by the friction of the stones rubbing together and the weight of water on top of same, and there is not one piece in twenty which is any good to them unless it has a little shell in it, which is useful. Ducks require a great deal of shell material. They must have it in some form or other. Oyster shell is the best material they can have. Of course, this is for producing the shell of the egg. Ducks should always have a good supply of oyster shell, as it not only helps to make the shell of the eggs, but also helps them to digest their food. Those who cannot get oyster shells can sometimes obtain cockle shells from the fishmonger. Shell, as well as grit, is of great importance to the health of the ducks.

Where there are only a few stock ducks kept, and they have the run of a field or farmyard, as a rule they find plenty of shells. They will often break and eat the shells of the snails. When they are well supplied with these things, which are really necessary, they will not pick up anything which is likely to injure them, but when they are kept short of grit they will swallow pins and needles and pieces of rusty nails, or any hard substance that they can get hold of. Instinct appears to show them that it is necessary for them to have something to masticate their food. It was thought a few years ago, that Aylesbury ducks could only be kept in that neighbourhood, on account of the grit which is found in abundance there. It is a kind of sea-shell and sand mixed up together, which of course is of a rough nature. This has been proved pure supposition, as they can be reared and kept in perfect health elsewhere, and there are just as fine ducks bred in the north as in the south, or anywhere in the United Kingdom, as long as they are treated properly. Of course by this I do not mean to condemn Aylesbury grit. It used to be said that if Aylesbury ducks were taken from that part the ducks could not be kept the right colour, because of the Aylesbury grit, as it caused their beaks to be the beautiful flesh colour which the bills of all prize Aylesburies should be. Now I find it is nothing of the kind. The beaks of the Aylesbury ducks become just as much discoloured in that neighbourhood as anywhere else. It is really the sun which gives the yellow tint to their bills, and wherever Aylesbury ducks are kept out in the sun too much they always get this tint more or less. It does their beaks good when they are allowed to swim in a running stream, as they find a lot of grit and sand, in which they plunge their beaks, and this keeps them clean. When the bills of ducks are continually covered with dirt, of course it soon spoils the beauty of them. If there is one thing the Aylesbury people lack especially it is sharp grit. Now the gizzard of a duck is small, in proportion to the size of the bird and the quantity of food it consumes; therefore they should be supplied with a good material to grind up that food while passing through the gizzard. If they have inferior material, such as grit with no sharp edges, or stones or shells which are too large, they only fill up the gizzard. I have frequently found large round stones in the gizzard when I have had to open them for post-mortem. Those of my readers who have made a study of this subject will have found that the inlet to the gizzard is more than double the size of the outlet, and coupled with this the duck has a large swallow. When they are kept short of grit they will often take hold of large round stones, or pieces of coal, and swallow them, and, as the inlet of the gizzard is large, these big pieces are admitted, and have to remain there, in some cases for months, until they are worn sufficiently small to pass out of the gizzard. In many cases this causes the duck pain, and sometimes death. When a duck has anything large and hard in the gizzard that it cannot pass, it will often waste away until it becomes a skeleton, because, when the large stones get over the passage leading out of the gizzard, it prevents the food from passing out of the gizzard after it is ground up. For instance, when a duck has been eating corn, especially oats or barley, the husks form a hard substance, and the gizzard becomes entirely blocked, and the ducks are unable to pass anything out of the gizzard, except liquid, and that gives them diarrhœa very badly, frequently ending in death.

HOW TO MAKE DUCK PONDS.

Duck Ponds should be substantially made—Cement, Concrete, and Ornamental Duck Ponds—Cheap Substitutes for a properly constructed Pond—Ponds for Stock Ducks: Fertile Eggs or Unfertile Eggs.

MANY people have a good space at their command, and yet have no pond for a pen of breeding ducks. These ponds can be made either round or square, but it depends very much on where they are to be made, and the number of ducks it is proposed to let swim in them, and also upon the soil. I will endeavour to explain briefly the way to make a good substantial pond, either round or square. I have known duck ponds which have been made that require repairing every time there is a sharp frost, which cracks the cement, letting the water through. Now this pond, though cheap in the first instance, is very expensive and unpleasant in the long run. When it freezes sharp the earth swells, and in this way raises the cement up and causes it to crack. These cheap ponds are made by simply cutting the earth basin shape, and then putting a layer of cement about an inch thick. This, of course, is not really sufficient, and having them made in this way is only a waste of money. A basin-shaped or round pond is a very good one for ducks, as the sides are slightly rough, besides slanting a little, so that the ducks can easily walk up out of the water. When such a pond is made it should have some flint stones or brick rubbish put at the bottom—when I say the bottom, I mean underneath the cement. These stones, or brick rubbish, should be at the least three or four inches thick all over, and the cement should be poured in among this rubbish; then it should all be faced over with cement, so that it is done at the same time, and will not chip off in little pieces. The cement on the top of the stones only requires to be about an inch thick, as the stones, or brick rubbish, really form a solid foundation. It is well to have a lead or iron pipe laid down, so that one end of the latter is a little below the bottom of the pond. It should, of course, come about the centre. A plug should be made to fit in the end of the pipe, so that when the plug is pulled out the water can run away quite easily; and, if the pipe is, say, a 4 or 6-inch, any sediment which may be lodged on the bottom of the pond will not be likely to choke the pipe up. There should then be another pipe forced into the 6-inch pipe, so that it stands up in the pond, so as to act as an overflow pipe. Of course this latter must be fixed in a larger pipe, so that the plug can be taken out and put in easily, and leave no chance of leakage at all. Should this smaller pipe become stopped up at all, it can easily be cleared by forcing a cane down and working whatever is the cause of the stoppage into the larger pipe. I might just mention that it is well to have the larger pipe laid down rather on the slant; then the water runs through more freely.

Whenever the pond is cleared out it should be brushed up well with a broom while the water is running through. Of course, those who make this kind of pond and have water laid on, should have a small pipe and tap to fill the pond; but in any other case it will have to be filled with buckets, or better still, if a little wooden trough can be let into the pond from some of the outbuildings, then the pond can be replenished by the heavy downfalls of rain. For instance, a wet day would be the time to re-fill it. Pull the plug up and well brush round the pond with a hard broom; this only wants doing occasionally, when there is an offensive smell arising from the water. A pond of this kind can be made any size. It is in the shape of a round basin, only with the sides more slanting, and can either be made one yard across or twenty, according to the number of ducks required to swim in it, and this pond will look ornamental as well.

The pond need not be very deep for ducks; 3 feet in the deepest place is quite deep enough. It is rather easier to make a square pond; there is less work in it. A pond of about 6 feet by 4 feet is large enough for ten ducks. If there are likely to be more it is best to make it larger. A pond of this size is large enough for two pens of ducks if they are of two different breeds. As soon as the pond is ready for use it can be divided by wire netting, and then several breeds can be kept; but the wire must be put down some distance under the water as well as above, so that the ducks cannot go down under the water and come up in their wrong pens. Some people prefer a deep pond; but a shallow one is much to be preferred in one respect, viz., it can be drained out better and takes so much less water to fill it; about 1 foot 6 inches or 2 feet is really as deep as ducks require. Whatever size the pond is to be, it should be cut or made about 6 inches larger than it is intended to be. If it is to be 4 feet wide and 6 feet long it should have the outsides a little on the slant, viz., the top should be cut 6 feet 6 inches, and the bottom 6 feet; this is the measurement from end to end. The sides should be cut in the same way, 4 feet 6 inches at the top, and 4 feet at the bottom, this to be measured from side to side. When made in this way it is a little on the slant, and the sides are not so liable to fall in when any weight or pressure is brought to bear upon them, such as a horse, cow, or cart wheel. When the pond is cut in this way the bottom should be concreted 3 or 4 inches thick. The concrete can be made with lime and gravel, or any kind of stones, only they should not be too large. It is well to break the largest, as they bind so much better. Any kind of brick rubbish, or flints, will answer the same purpose. Cement can be used, it is more substantial, although it is more expensive; but the concrete makes a very good pond. A little sand or road scrapings should be mixed with the lime, as this acts as a binder. If cement is used for the bottom, 2½ inches will be thick enough; if lime,

3½ inches. A pipe of some kind should be let into the bottom, so that the water can be easily drained out when required. It is well to have an iron pipe laid in so that the bottom part of the pipe is 1 inch lower than the bottom of the pond. Have a good-sized pipe, either 4 or 6 inch; then it does not get blocked up, and the water, with the sediment, will all run out easily. It should be brushed and stirred up with a broom while it is running out. If the water has to run far, a clay pipe can be used; this is much cheaper. The pipe should be a little on the slant, then the latter keeps clean, as the force of water clears it. There should also be an overflow pipe, same as in the round pond. The posts which help to form the frame-work should be put in before the bottom concrete is laid; they do not require putting in deep, as there must be some cross pieces to keep the sides firm when the concrete is put in. The boards should be placed 3 inches from the sides of the pond, according to the thickness the walls are required to be; then the concrete should be put in all round. These boards form a mould. Let it remain a few days before the boards are taken down, then it will have set hard. When the boards are taken down and the posts drawn out, the holes should be filled up with good cement. After the boards are taken away, the surface of the concrete is rough. It should not be left thus, as the dirt clings to the rough surface, so that it cannot be cleansed. It is best to plaster it over with clear cement, thus giving it a clear and hard surface, which will stand both water and frost, and can easily be cleaned. The pond can be filled by means of an India-rubber pipe, fastened to a tap. A better way of course is to have the water laid on so that it can run into the pond. In some places there is no water laid on. In this case it is well to connect, so that the latter from the outbuildings runs into the pond. Then it requires to be emptied occasionally; this cannot be well done by a tap, as the grit gets in and spoils it, so that it leaks, whether it is wood or brass. I find the best and easiest plan for this purpose is to have a piece of wood about a yard long, a little larger than the pipe, quite round, sharpened a little at the end, then wrap a piece of strong linen or flannel round it; just knock it in gently, so that it is firm. This usually acts splendidly, and it is no trouble to repair, only to put another cloth round it occasionally.

Ducks require both sand and grit, and it should not be put into the pond; if so, it will get washed away when the pond is cleaned out. A trough can be placed at the side, or it can be so constructed that the trough is in the pond, just underneath the surface of the water. This can be done by putting some iron rests in the concrete wall for the trough to rest on. When done in this way it is very convenient for the ducks, as corn can be put in as well as grit. If fowls are not allowed to run with the ducks, the pond may be made so that there is a sloping place for the ducks to walk down into it. If fowls are kept with them it is best to have a little plank ladder (a board with little pieces of lath nailed across it). This prevents the ducks from slipping. It should reach from 6 to 8 inches in the water. Then the ducks will get on it when they want to leave the pond. If an ornamental duck pond is required, it should be made a little deeper, and not filled up to the top. In this way, tree roots, old stumps, and pieces of rough wood, can be embedded in the concrete in a rustic manner, and then some large common ferns planted, such as brake ferns and nest ferns. Both these varieties grow up very large, and spread out, partly covering the rustic work. A few rushes may also be planted. These are very hardy, and grow fast. It gives the pond quite an ornamental appearance, and can be done in one year. If planted before March they will grow up beautifully the same summer. Those who have only a drake and three or four ducks, with a very small place to keep them in, and it is not convenient to make a pond of cement or concrete, which costs time and money, can sink a round tub into the ground, and arrange it so that there is a plug-hole to let the dirty water off. A tub eighteen inches or two feet across, the latter the best, does very well for one pen of ducks. A paraffin barrel cut in two through the middle answers nicely. No matter what kind it is so long as it holds water. Some people use a galvanized tank or an old lead cistern. Occasionally things of this kind can be bought up cheap at a builder's yard, or a second-hand shop in the town. Where one has not the means to make proper duck ponds, a cheap substitute such as I have just mentioned will answer the purpose very well. Under any circumstances I do not advise tubs or ponds for young ducks till they are full grown, unless just to wash in them; even then they are better without one. These ponds are for the stock ducks which are kept for breeding, as it is natural for ducks to be in the water when they are mating; and the eggs, as a rule, are more fertile. When ducks have never been allowed to go in the water, the eggs are as fertile without as with a pond, but where they have been accustomed to have a pond to mate in, and then are moved to a fresh place without one, the first few batches of eggs will probably be unfertilised. I know many large breeders who do not allow their stock ducks to have a pond at all; rearing them from the shell with only sufficient water to drink, and their eggs are very fertile.

DISEASES.

Ducks not subject to many ailments—Treatment for cases of consumption and oviduct displacement—Roup, staggers, and loin weakness in ducks: how to deal with.

DUCKS are not so subject to diseases as poultry. There is very little hurt comes to stock ducks, except being egg-bound, ruptured, and occasionally there are cases of liver disease. The latter very seldom appears amongst ducks when the stock comes from healthy birds. Therefore chapters on different diseases will not be required as with fowls. When they have liver disease they usually go lame in one leg, and get very thin, but eat well at the same time. When these symptoms are noticed it is far better to kill the duck off at once. Do not attempt to doctor such birds; but do not make a mistake and kill a bird simply because it has got lamed in some way. A duck will occasionally go in consumption. When they have a touch of this disease it will be noticed that they do not care about eating their soft food, but will usually eat corn (particularly Indian) very ravenously, and at the same time become light in weight

In all cases, when this disease is seen in a bird, it is best to kill it at once. Ducks may get a little out of order. Sometimes they eat things which disagree with them, and they do not appear to have any appetite for a few days. When this is the case, give them a teaspoonful and a half of salad oil each, and a teaspoonful of roup powder to every four full-grown birds. If this is done it will usually put the ducks all right in a few days. Sometimes, when ducks are in full lay, they have the cramp. In this case they should be kept in a small, dry place by themselves, and not be allowed to run with the drake. The food should be given them warm, and they are generally all right again in about three days. It is not very often that ducks are egg-bound, if they are fed properly; but the symptoms are as follows:—The duck goes to the nest a good deal, and stays there a long time—sometimes for hours—and she will be noticed to be a little down behind. When these symptoms are observed a little sweet oil should be put in the oviduct with a feather. This will often help them to pass the eggs. Sometimes, when a duck lays a very large egg, especially a young duck, the oviduct comes down. When this is the case do not attempt to put it back, unless it is a very valuable duck, but kill at once, as the flesh is good to eat. Should it be a very valuable bird, then it is worth while to try the experiment. Oil the oviduct well. Often it may be a little soiled by the duck running about; then it should be washed with a little warm water, and wiped perfectly dry with a nice soft cloth; and afterwards covered with sweet oil. Wrap the cloth round the duck's body and wings, allowing the legs to be outside;

then hang her in a sling, so that the head and shoulders are about three inches lower than the end of abdomen. A roller-towel is a good thing to hang her in. Let her hang like this for from six to twelve hours, and, as a rule, the oviduct will work back again to its proper place without forcing the bird to strain herself in trying to get it back. If they are allowed to run about at once the oviduct will force its way out again, and cause the duck to strain. But when she is left perfectly quiet, hanging with the abdomen so much higher than the other part of the body, eight out of every ten can be cured in this way. They should always be kept short of food for a few days after this operation has been performed, to keep them from laying eggs before the oviduct gets thoroughly back in its proper place and all inflammation has passed away. Sometimes ducks will have a slight touch of roup. The only symptom which can be seen to distinguish this by is a little foam and dry matter round the eye—they look as though their eyes were dirty. When this is the case they should have their eyes bathed and all the discharge washed off. In some instances this makes their eyes a little sore; and it is well to put just a little vaseline round the eyelids, and at the same time give a little roup powder. When this disease appears it will often attack every bird in the pen; but, when the roup powder is used, in many cases no other birds will show any symptoms of it, if taken in time. One heaped-up teaspoonful of the powder to every four or five full-grown ducks, put in their soft food in the morning, is about the quantity that should be used. This disease seldom attacks stock ducks, but young ducklings are more subject to it. If they are allowed to sit in a draughty house or coop, or are let run in the hot sun, and then get a sudden chill, they will often get the roup, and are subject to it quite as much as chickens are. It frequently sends the young ducklings blind. What I mean by this is that the discharge from the eye collects and gums the eyelashes together. When this occurs it stops the growth of the young ducklings very considerably.

The eye should always be bathed in warm water as recommended for the old ducks, and roup powder given, varying in quantity according to the size of the ducklings. It will not hurt them if they have a little too much, as it is not injurious, the proper quantity is one heaped-up teaspoonful for every eight or ten young ducklings at about three weeks old, once a day. There are hundreds of young ducklings die every year with roup, and yet their owners have not the slightest idea what is wrong with them, and therefore never give them anything to relieve them. I am sure of this, because I so often see it as I am travelling, and I frequently have a large number of them sent to me for post-mortem examination. Young ducklings when they are from three to fifteen days old are rather subject to inflammation of the lungs. When they are allowed to run in very cold water, after the hen has been brooding them, or they have come out of a warm house, the sudden change causes this complaint. They should be kept at one temperature as much as possible, and in all cases the sleeping coops or houses should be kept very dry, as " prevention is better than cure." Occasionally young ducklings will have what most people call the staggers. This is to say, they get very giddy, and they often fall on their backs, and cannot get up again. This is caused by the sun striking so hotly on their brain, or being kept too thickly in a house. The only thing I have found useful in this case is to put their heads under a tap of cold water, or a pump, and let the water run for about five minutes. They can often be brought round by serving them in this way. Where ducklings are kept very thickly together, there is occasionally one or two among the number which appear to swell very much. Their necks will swell to almost the size of their body, and the latter will swell to double the size it ought to be. This is wind or air which gets between the flesh and the skin. Whenever they are noticed like this, a small, very sharp penknife, or lancet, should be used, just to cut the skin where the swelling is forming; then squeeze the enlargement gently, and the confined air will escape, just like letting air out of a bladder. Sometimes they will swell again, but, as a rule, when it is taken in time no more is seen of it. Young ducklings are also very subject to weakness of the loins, and when this is the case they should be kept very quiet, and put with younger ducks, or have a place by themselves, and have a little bonemeal added to their food, because those of their own size, when they are let out, run much faster and tread upon them. This frightens them very much, and makes them worse. It is better to have them kept in a small quiet place, where it is nice and dry, and fed well; then they will usually recover, and at the same time often grow quite as fast as the others.

But they should not be allowed to be driven or run fast, as this frightens them, and causes them to be much worse. These are just a few ailments which young ducklings are subject to, but they are really very little trouble in comparison to chickens. There is one thing I must mention. Ducklings do not show symptons of being ill so much as chickens. If a young duckling has inflammation of the lungs, it will run and slip about with the others, and though it may not feed with them, it will be running about among them, so that the attendant has a difficulty in detecting whether they eat or not, unless he is watching very closely. Ducklings are often found dead before they are noticed to have anything the matter with them, and when they are opened they have inflammation or congestion of the lungs very badly.

SUPPLEMENTARY CHAPTER.

Ducklings for the multitude—The plan impracticable—Present Position—Eggs and Ducklings: Counsel and Advice.

IN a former edition of the Duck Book I explained the circumstances under which I was led to offer to supply ducklings to those who felt they were unable to undertake the hatching of broods for themselves. After making arrangements for a large number—it was found that the ducklings being often some days on the journey from one place to another suffered fearfully from thirst, and a large number of them died.

I found it therefore an impracticable thing and abandoned it, because I could see it was never likely to be a success. I shall be glad to supply sittings of eggs of all the varieties of ducks, as well as ducklings of all kinds, at an age when, having got their feathers, it is not so risky to send them on a journey.

There is no reason why every family, having the very smallest accommodation, should not rear a brood of ducks for their own consumption, and thus enjoy the luxury of a roast duck, which perhaps would not come within their reach in any other way.

I have treated fully upon every phase of duck rearing in the chapters of this book, so as to enable those who have been in doubt hitherto as to the best way to set about the business to do so with security.

WHITE INDIAN RUNNER DUCK.

WHITE INDIAN RUNNER DRAKE.

INDIAN RUNNER DRAKE.

ROUEN-AYLESBURY CROSS.

BLACK EAST INDIAN DUCK AND DRAKE.

MUSCOVY DUCK AND DRAKE.

CAYUGA DUCK AND DRAKE.

PEKIN DUCK AND DRAKE.

ROUEN DUCK AND DRAKE.

AYLESBURY DUCK.

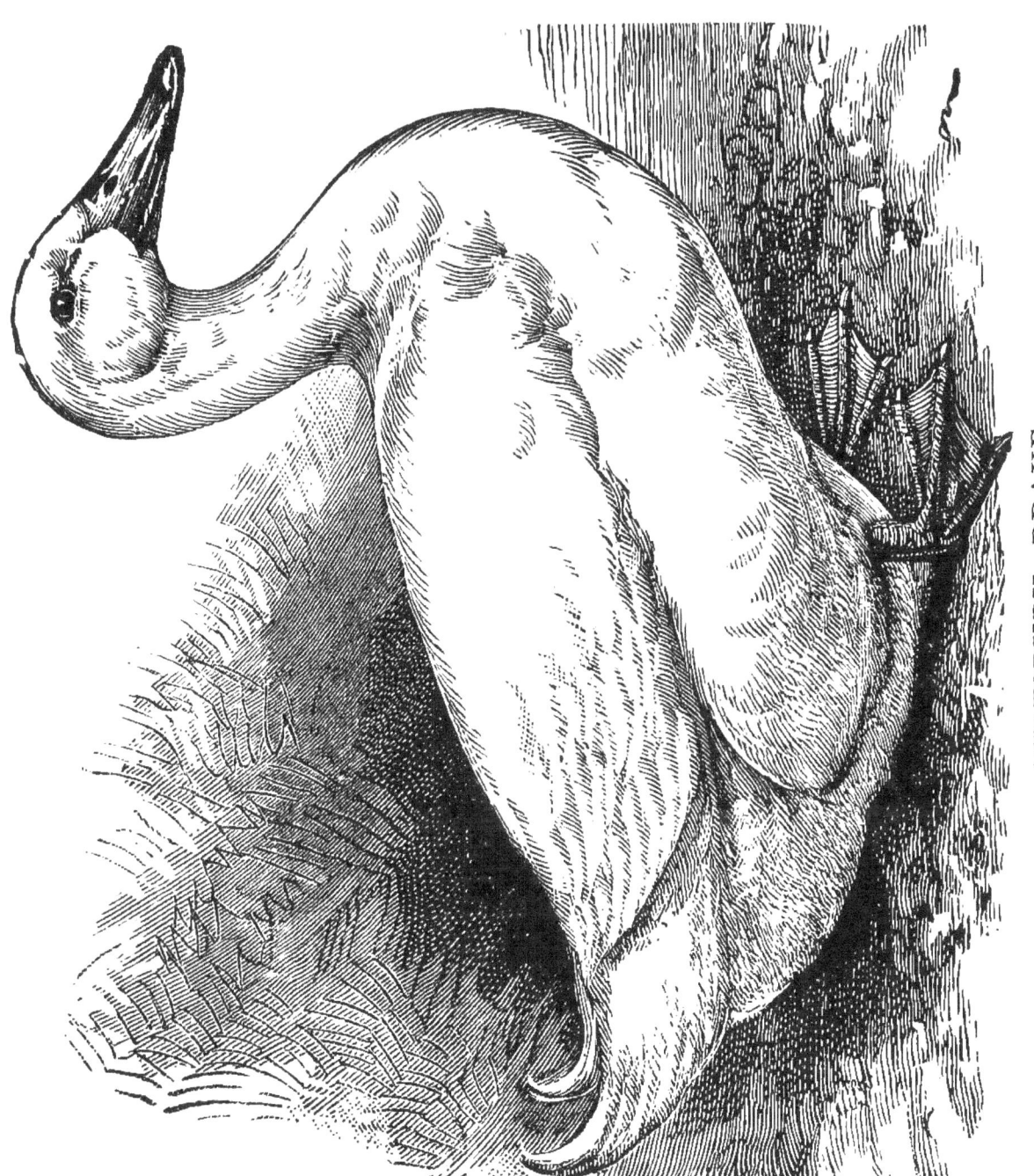

AYLESBURY DRAKE.

ADVERTISEMENTS.

OLD CALABAR

DOG BISCUITS

AND

POULTRY FOOD

WITH

MEAT and VEGETABLES.

PRICE 20/- PER CWT.

ADDRESS:—

The **OLD CALABAR** Biscuit Co., Ltd.,

28, Chapel St., LIVERPOOL.

ADVERTISEMENTS.

LECTURES.

—:o:—

W. COOK gives Lectures on Practical Poultry Keeping and Management, throughout the Country. His object is, as fully expressed in his book, to draw the attention of the labouring classes to this easily-attained means of adding to their income. Write for terms. Special arrangements made with Secretaries of Technical Education Committees or Poultry Societies for Courses of Lectures.

ADVICE & CONSULTATION.

—:o:—

W. COOK & SON give information free to all poultry keepers and duck rearers, on the management of poultry, also answer any question by stamped and addressed envelope. All communications concerning this should be addressed to "Orpington House," St. Mary Cray, Kent. They also travel to all parts of the United Kingdom for the purpose of planning-out poultry farms and runs, mating breeding birds, &c., for the nominal sum of £1 1s. and travelling expenses.

NOTICE TO VISITORS.

—:o:—

The Breeding Pens and Stock Birds are open for inspection any weekday. Mr. COOK is at home on Wednesdays from 1.30 p.m. till evening, to supply every necessary information concerning the birds, and give advice. For the benefit of the working classes, the place is open for inspection on Bank Holidays. "Orpington House," St. Mary Cray, is 2½ miles from Orpington Station on the South-Eastern Railway, and a little over a mile from St. Mary Cray Station on the L. C. & D. Railway, and 2 miles from Swanley Junction on the same line. Anyone who cannot conveniently walk can be met by trap at either station by appointment. Address all Telegrams:—"Cook, St. Mary Cray." Porterage on Telegrams to St. Mary Cray, 6d., which must be prepaid. Address:—Orpington House, St. Mary Cray, Kent; and Queen's Head Yard, 105, Borough, London, S.E. (3 minutes' walk from London Bridge Station).

WILLIAM COOK & SON,

Orpington House, St. Mary Cray, Kent.

ADVERTISEMENTS.

NINTH EDITION,
RE-WRITTEN AND FULLY REVISED TO DATE.

THE PRACTICAL
POULTRY BREEDER
AND
FEEDER;
OR,
HOW TO MAKE POULTRY PAY.
By W. COOK.

PRICE 2s. 6d.; POST FREE, 2s. 10½d.

This Book, which is practical in every detail and based on the experience of the writer, has the largest circulation of any book of the kind ever published.

It deals practically with all subjects. The chapters on the laying qualities of fowls, diseases, and preparing birds for the show pen being especially useful. There are also chapters on older breeds as well as those which have more recently been brought out, and no poultry keeper should be without one for reference. (Refer to press reviews).

W. COOK & SON, Orpington House, St. Mary Cray, Kent.

ADVERTISEMENTS.

OPINIONS OF THE PRESS.

Practical Poultry Breeder and Feeder; or, How to Make Poultry Pay (by William Cook).—The idea of likening poultry unto machines for converting waste and worthless matter into very good and profitable delicacies, is a happy one, and the author explains it very thoroughly. With little labour and attention, fowls may be kept so as to yield a good return: but there are conditions which must be observed, and these are simply and plainly laid down by the author, who is the most careful instructor we have met with for a very long time.—*Daily Chronicle.*

Practical Poultry Breeder and Feeder; or, How to Make Poultry Pay (by William Cook).—We are perfectly sure none who follow the plain instructions given will fail to keep poultry at a profit.—*Glasgow News.*

Cook's Practical Poultry Breeder and Feeder.—Mr. Cook says that poultry may be likened to machines for converting waste and worthless things into good and profitable delicacies. Much good poultry food is, no doubt, thrown in the hog-tub, or otherwise wasted. How to turn many things to account may be found in Mr. Cook's valuable manual, which is full of practical knowledge of all kinds of fowls and their management. Poultry breeding and keeping not only *may be*, but *it is* profitable when carried out on the system recommended by Mr. Cook.—*Land and Water.*

How to Make Poultry Pay.—This is the title of a practical work on poultry breeding and feeding, by Mr. William Cook. It is especially adapted for cottagers, or those having limited accommodation for keeping poultry, and the author has been successful in his endeavour to impart plain and practical information, which will be of service to the amateur poultry breeder, and enable him to make it a profitable pursuit. —*The North British Agriculturist.*

ADVERTISEMENTS.

Mr. Cook in his useful little book, *How to Make Poultry Pay*, remarks that the number of eggs annually imported by this country is about 750 millions, worth (say) £2,400,000. As is generally known, the majority of these eggs come over from France, where they are produced by cottagers and farmers, nearly all of whom keep fowls, and make them pay well. Mr. Cook thinks that if our cottagers and farmers would only devote themselves to a little practical study of fowls and their rearing, at least one half this sum of money could be kept in this country. A friend who followed Mr. Cook's sensible advice was able to increase his store of eggs from 400 to nearly 800, without, at the same time, adding to the number of his fowls.—*Society*.

How to Make Poultry Pay (by William Cook).—Mr. Cook points out so many facts concerning the numerous errors universally made, either through ignorance or prejudice, about poultry, its rearing and breeding, that the little manual deserves to be widely dispersed. It has often been said that the English working-classes might be much better off than they are if they only knew how to take advantage of things, as do the French, who in reality are exceedingly poor, but at the same time very frugal, and admirable in their perfect knowledge of domestic economy, often knowing how to live comfortably on what their English fellow labourers throw away. Mr. Cook's book, however, has a wider scope than that of teaching *poor people* how to keep poultry. It addresses itself equally to the rich, and so practical are the hints it contains that one gentleman by following them managed to increased his store of eggs in one year from 1,800 to 2,300, and yet he did not add to the number of his fowls. He simply punctually obeyed Mr. Cook's rules for dieting his poultry, and the result was such as greatly to surprise and delight him.—*The Morning Post*.

Poultry Breeding and Feeding.—So much has been written of late years in connection with the subject of this little work, that one feels disposed to doubt whether there be anything that is to be told. Mr. Cook, however, takes up the subject in a somewhat different spirit to that of most writers.—*Journal of Horticulture*.

The Practical Poultry Breeder and Feeder. By W. Cook.—The book abounds with useful information requisite for the management of poultry with a view to profit as well as pleasure, the information being explained in a thoroughly practical and simple manner.—*Norwood Review*.

Practical Poultry Breeder and Feeder; or How to Make Poultry Pay. By William Cook. The fifth edition, re-written and revised to date.— This would enable anyone who had little or no idea of poultry "to keep fowels and make them pay well, thus combining pleasure and profit, both in town and country." (Queen's Head Yard, 105, Borough, London.) —*Newcastle Daily Chronicle*.

ADVERTISEMENTS.

Practical Poultry Breeder and Feeder ; or How to Make Poultry Pay. By William Cook. Fifth edition, re-written and revised to date.—A very complete and plainly-written manual for those who wish to keep poultry and to combine in so doing both pleasure and profit.—*Literary World.*

A fifth edition of *The Practical Poultry Breeder and Feeder*, a useful and well illustrated manual for those who wish to know the ways of fowls and profit by them, has come from Mr. E. W. Allen, London. The author of the work is Mr. William Cook, who has increased its value in this edition by a thorough revision and addition of some new matter.—*The Scotsman.*

Practical Poultry Breeder.—Many people would like to know "how to make poultry pay;" still more, perhaps, would like to know how to manage fowls for amusement or domestic purposes. Such persons may be glad to hear of a little book under the above title written by WILLIAM COOK, an expert in poultry raising.—*Gardeners' Chronicle.*

Poultry breeders should welcome the appearance of a new edition of Mr. William Cook's *Practical Poultry Breeder and Feeder* (E. W. Allen, Ave Maria Lane), as the many valuable directions on management, feeding, &c., contained in the work cannot fail to prove serviceable to all who keep fowls, whether for pleasure or profit.—*Graphic.*

Poultry Breeder and Feeder. Published by the author, William Cook, at the Queen's Head Yard, 105, Borough, London, S.E.—Those who had the pleasure of perusing Mr. Cook's valuable work when it made its appearance a few years ago, will not be surprised to learn that it has run into the fifth edition. But though the present issue is styled a fifth edition, it is to all intents and purposes a new book. It has been re-written and brought fully up to date. Everything of special value that appeared in former editions has been retained, and a variety of new matter, rendered necessary by the altered state of things under which we live, has been added. The book in its new form, should therefore be of exceptional interest to breeders and rearers on poultry, whether for fancy or for the market. The growing importance of the latter point has not been over-looked by Mr. Cook. For the guidance of those who would give attention to this matter the author imparts a deal of useful practical information which should assist the poultry-raiser in making it

ADVERTISEMENTS.

a profitable pursuit. The most suitable class of houses, the best system of breeding, feeding, and rearing, are all dealt with in detail; while several valuable hints are given as to the selection of the best pure breeds for crossing. The best poultry for egg-producing and table purposes are indicated, and altogether the work should commend itself to all who are interested in the question of poultry-raising, which is receiving increased attention every year. Mr. Cook's book comes up to its title in a much fuller degree than any other work on the same subject with which we are acquainted.—*North British Agriculturist.*

Cook's Poultry Breeder and Feeder; or, How to Make Fowls Pay.—Mr. Cook's success not only as a writer about the history of poultry and the points of the various breeds, but as an instructor and example how to manage breeds to make them pay, is generally acknowledged. As this is a fifth edition it is plain that the public only require to be told that the book is "in print" once more.—*Live Stock Journal.*

Practical Poultry Breeder and Feeder; or, How to Make Fowls Pay. By William Cook. Fifth edition revised to date. London: Published by the author.—The writer of this handbook makes hens and eggs his business, farming them himself at "Orpington House," St. Mary Cray, and going up and down the country to help others in doing the same by lectures and advice. So far as we can judge the book seems sensible and useful.—*Liverpool Mercury.*

Poultry Feeder and Breeder.—There is no work better known and appreciated than this work of Mr. Cook's, the fifth edition of which, re-written and revised to date, we now welcome.—*Bell's Weekly Messenger.*

The Practical Poultry Breeder and Feeder; or, How to Make Poultry Pay. By W. Cook.—Unlike many writers on Poultry and their management, Mr. Cook is a large breeder himself, and has spent many years in making experiments with most kinds of poultry. His writings should be more valuable on that accout. The opening chapters of his work have the most interest for ordinary poultry-keepers, as they contain general directions for breeding and feeding.—*Farm and Home.*

W. Cook and Son, Orpington House, St. Mary Cray.

ADVERTISEMENTS.

THE POULTRY JOURNAL:

How to make Poultry Pay.

EDITED BY WILLIAM COOK.

PUBLISHED BY

E. W. ALLEN, 4, Ave Maria Lane, E.C.

THIS is a Monthly Paper, and is the only monthly journal in England which is devoted entirely to Poultry. In each number there is a chapter of Hints for the current month, according to the season of the year, showing how to manage both the old and young stock, &c. Also short chapters on ducks, turkeys, and geese, and their management, &c., when kept in small runs. Questions are answered through the columns of this paper, and also free by post by enclosing a stamped and addressed envelope. Post-mortem Examinations are made on all kinds of Poultry, for the nominal sum of 1s. each. All specimens for examination to be sent, carriage paid, to 105, Borough, London, S.E. The Reports appear in the Monthly Journal, and in cases of urgency, if a stamped and addressed envelope is enclosed, they are answered by post. In cases of contagious disease, a letter of instruction is sent free of any other charge. Specimen copy of the journal sent free of charge.

ORPINGTON HOUSE, ST. MARY CRAY.

PRICE, TWOPENCE PER COPY.

Postal Subscription: Three Months, 7½d.; Six Months, 1s. 3d.; Twelve Months, 2s. 6d., payable in advance.

ADVERTISEMENTS.

W. COOK & SON

SUPPLY

EGGS FOR HATCHING

From over Forty Varieties of Poultry, also many Breeds of Ducks, Turkeys and Geese.

SEND FOR

Full Current Price List

POST FREE,

As the Eggs vary in price so much according to time of year.

ADDRESS:—

W. COOK & SON,

Orpington House, St. Mary Cray.

(No connection with J. W. Cook, of Lincoln.)

W. COOK & SON'S POULTRY POWDERS.

These Powders are an invaluable composition for Poultry under all circumstances. They are prepared especially to act upon every organ of the body, being stimulating, strengthening, and warmth-giving—in fact, they counteract many diseases Poultry are subject to, improve their appearance by imparting a gloss and beauty to the plumage, and keep the fowls in good health by preventing colds and hardening the birds against the severe and constant changes they are subject to in this climate.

They are especially useful to birds while moulting, when there is a great strain upon the system in the growth of the young feathers, and they are down in condition and need something to help them. They are also useful in cases where fowls mope about and do not care for their food, being a little out of sorts. The powders will be found most beneficial by acting upon the liver, and bringing the birds on to full lay. Those who use them are seldom without eggs all the winter months. They are used very largely and have proved a great boon to poultry-keepers. They do not over-stimulate the fowls and leave them weak, like most other tonics do. They strengthen every organ of the body, and can be discontinued at any time without any injury to the fowls. I have used them for more than sixteen years, from August to April, about four or five times a week; if the weather is severe I use them every day. Many people have used them all through the summer of the past few years with excellent results; they do not injure the birds in the least or wear them out sooner, as my customers testify. I have not been without eggs for more than sixteen years, even in the most severe weather. The same Powders are used for bringing up young chickens, turkeys and pheasants, and this year they have been used with great advantage for young ducks; they have a good effect on all young poultry, assisting them in their growth, getting their feathers, and giving them health and vigour.

The quantity to be used is a full teaspoonful for eight full-grown fowls, and chickens proportionate to age, given from three to five times a week, with the morning meal of soft food; it is best to mix the Powder in the dry meal previous to adding the water, or it can be mixed in any kind of soft food. When the fowls are in full lay, or the weather mild, the Powder may be omitted for a week or so. A change does them good. I cannot state exactly the time to give it to young chickens, but early-hatched ones require it oftener than later-hatched, as the former suffer much from cramp, cold, &c., according to the weather. The use of it must depend upon the feeder's judgment.

These Powders are sent to all parts of the world. They help the fowls to produce eggs in the coldest weather, and also when kept in close confinement eggs are produced in abundance. Where many did not get an egg for three months together, since using the powders they are never without them.

Sold in tins, post free, 8½d., 1s. 3d., 2s. 4½d., 5s., and 12s. tin for 10s.; also supplied in larger quantities in linen bags at a reduced rate. Cash to accompany all orders.

W. COOK & SON'S ROUP POWDERS.

PRESCRIPTIONS FOR MAKING UP THE PILLS.

FOR THREE HENS.—One heaped-up tea-spoonful of the Powder, two ditto of flour, two ditto of oatmeal, two ditto middlings or any kind of meal, with a small piece of fat of some kind, about the size of a walnut; mix with a little warm water into a paste, so that it does not stick to the fingers. The pills should be made about the size of the little finger to the second joint. Give two pills night and morning. If the fowls have the disease very badly, a little extra powder can be used without injuring the fowl in any way. Though very strong, it is not poisonous. If the invalids cannot eat, they should have something nourishing, such as bread and milk or stewed linseed, given warm. Always let them have as much food as they can digest. When the fowls first show symptons of roup they ought to have a teaspoonful of castor oil given to them, and half a teaspoonful of glycerine; even when they show signs of a cold they can have this given them. Isolate affected birds, and add camphor to all drinking water. Give the unaffected birds the Roup Powders in their morning meal—one heaped-up teaspoonful to ten fowls. This will often stop the malady from going further. On a cold or damp day a little of it is most valuable. When fowls are going on a journey, a pill or two will often prevent them from catching cold. Many exhibitors use it and find it most beneficial. A preventive is better than a cure. It has saved the lives of thousands of fowls all over the world. In many cases it has cured them when all other advertised remedies have failed. If fowls are suffering from lowness, or their liver is out of order, it soon puts them right and brings a bright lustre on their plumage, which improves them very much for the show pen. When a fowl has a rattling in her throat and difficulty in drawing her breath, give her a teaspoonful of glycerine, and when convenient stew some linseed and give from six to eight teaspoonfuls warm. Keep the affected birds on straw or moss peat; the latter is much the best. When the birds have swollen eyes bathe them in milk and water with a little camphor in it. Always wipe the face and eyes dry; if not, they catch a fresh cold. When a fowl has a thick discharge, called mucus, which corrodes round the tongue and throat, use the lotion, 9d. per bottle, per post 10½d. Directions for use of same:— Take a feather, which dip in lotion, apply to the bird's mouth and throat, turn the feather well round the mouth—in this way it will bring much of the thick slime away. In bad cases it requires a second feather to repeat; then it is well to take a feather and dip in glycerine and also mop out mouth with that. This heals the wound. If this treatment is continued night and morning, the reward will be the bird's recovery.

Sold in tins, post free, 9d., 1s. 3d., 2s. 4½d., 5s., and 12s. tin for 10s.

W. Cook & Son, Orpington House, St. Mary Cray.

INSECT POWDER.

W. Cook and Son's Improved Insect Powder will destroy all insects on poultry, pigeons, cage birds, dogs and cats; also destroys black-beetles, and is used largely for household purposes; its use is indispensable in keeping the nest and sitting hen free from insects; is perfectly harmless. Should be freely used just before hatching, both on the hen and in the nest, as it is impossible for chickens to thrive when covered with vermin. Sold in tins, post free, 8d. and 1s. 3d.; or 5s. tins, carriage paid.

W. COOK & SON'S
FATTENING POWDERS.

These powders are very useful in assisting poultry to put on fat and to keep them in health at the same time; they give them a keen appetite, and assist digestion.

    For 12 Fowls, one dessert-spoonful three times a week
    ,, 10 Ducks, ,, ,,
    ,, 6 Turkeys ,, ,,

Sold in Tins, post free, 1s. 3d., 5s. and 10s.

SPECIAL QUOTATIONS FOR LARGER ORDERS.

W. COOK & SON'S
OINTMENT FOR SCALY LEGS,

Sold in 6d. and 1s. boxes. Post free 7½d. and 1s. 3d.

W. COOK & SON'S EMBROCATION.

Sold in 6d. and 1s. bottles. Free by post for 8d. and 1s. 3d.

W. Cook & Son, Orpington House, St. Mary Cray.

W. COOK & SON'S
PRICE LIST OF MEALS & CORN.

	1 cwt.		½-cwt.		¼-cwt.	
Biscuit Meal	17	0	8	10	4	8
Special ,,	15	0	7	9	4	0
General ,,	12	6	6	6	3	6
Bone ,,	14	0	7	0	3	9
Duck ,,	14	0	7	6	4	0
Fattening Meal for Ducks	14	0	7	6	4	0
Granulated Meat	21	0	10	6	5	6
Meat Dog Biscuits	14	0	7	6	4	0
Ground Oyster Shells	8	0	4	6	2	6
Flint Grit	12	0	7	6	4	0
,, Dust	8	0	4	6	2	6

Delivered free to any Railway Station in England. Half carriage paid on 1 cwt. bags to customers in Ireland and Scotland.

	Sack.		½-Sack.		Bush.	
Wheat (best)	19	0	9	6	5	0
Buckwheat (best French)	18	0	9	0	4	6
Barley	18	0	9	0	4	6
Maize (small round)	18	0	9	0	4	6
Dari	18	0	9	0	4	6

Groats (whole) extra quality, 19/6 per 112 lbs.; 10/- per 56 lbs.; 5/3 per 28 lbs.

Carriage paid ONLY within the delivery of Carter Paterson & Co.

Orders for CORN cannot be executed unless a remittance for Sacks or Bags accompany the Order.

Sacks charged 1s. 4d., 1-bushel bags, 6d., and allowed for when returned to London Warehouse, 105, Borough.

FLINT GRIT.

Sharp grit is most essential to all birds, as it is their only means of digesting their food and they cannot thrive properly without it. FLINT GRIT is the best material that can be used. W. Cook and Son supply FLINT GRIT broken in proper sizes for Fowls, Turkeys, Ducks, Pigeons, Chickens, and Cage Birds, and it is all broken by hand, as Grit broken by a machine loses the sharp edges it is so necessary to retain, and 28 lbs. of hand-broken flint will last Fowls longer than 112 lbs. of flint broken by machinery. FLINT DUST is also invaluable for laying Fowls and Ducks, it being strengthening to the egg organs.

W. Cook & Son recommend both the FLINT GRIT and the DUST to be mixed in the Fowls' soft food, waste thus being prevented. Fowls require about half a teaspoonful each weekly, of the Grit, and when in full lay three teaspoonfuls of the Dust daily to about 10 hens will be sufficient.

W. COOK & SON, Queen's Head Yard, 105, Boro', London and Orpington House, St. Mary Cray, Kent.

GERMAN MOSS PEAT LITTER.

—:o:—

Moss Peat instead of dust, ashes, or lime in the house, is the greatest boon to poultry keepers of anything that I know. It saves time, keeps the houses clean, and is in every way a comfort to the fowls themselves. If the houses are cleaned out four or five times a year, it is quite often enough, as the peat does away with all smells, an occasional stir-up being all that is required. When once used, a poultry keeper would not be without it for anything.

Sold in half-hundredweight bags	3/3
Three bags	9/-

Bag and free delivery per Carter Paterson included.

Bales, weighing from 2 to 3 cwt.	9/-
By the ton	45/-
„ half-ton	23/-

Purchasers must pay carriage on these quantities.

NEW EDITION.   NOW READY.

The Poultry Keeper's Account Book.
(W. COOK & SON'S).

The most complete method published. Price 1/-, post free 1/1½.

W. COOK & SON,
Queen's Head Yard, 105, Boro', London, S.E.

ADVERTISEMENTS.

W. COOK & SON

SUPPLY

STOCK BIRDS

From over Forty Different Varieties of Fowls, also Ducks, Turkeys and Geese for

Exhibition, Pure Breeding or Crossing Purposes.

———:o:———

INSPECTION OF OUR POULTRY FARM INVITED.

———:o:———

All Birds are sent on approval to any part of the country, on customer paying carriage both ways if Birds are returned.

Pure bred birds of all varieties can be supplied all the year round, either singly, in pens, mated, or large numbers. Hens or Pullets from 7/6, 8/6, 10/6, 12/6, 15/6, 21/-, 30/-, £2 2s. to £6 6s. each. Cocks or Cockerels, unrelated, 8/6, 10/6, 12/6, 15/6, 21/-, 30/-, £2 2s., £3 3s. to £8 8s. each. Birds for farm-yard purposes at low prices. Cross-bred Pullets, 5/6 and 6/6 each. Special prices given for Birds for Exhibition purposes. Foreign orders executed. Send for our full Price List of Birds, Eggs, and Poultry Specialities, Post Free.

W. COOK & SON,

ORPINGTON HOUSE, ST. MARY CRAY, KENT.

(No connection with J. W. Cook, of Lincoln.)

ADVERTISEMENTS.

NO POULTRY HOUSE

Is complete without one of the

PATENT AUTOMATIC

DOG PROOF SLIDES

FOR FOWLS' EXIT,

PRICE 10/-.

SOLE MAKER—

SAML. SUTCLIFFE, Queen's Road Saw Mills, Halifax.

Also Maker of good substantial Cheap Duck Houses, Fowl Houses, Chicken Coops, Dog Kennels, &c.

Wire Netting, Roofing Felt, &c., at cheap rates.

CATALOGUES FREE ON APPLICATION.   A TRIAL ORDER SOLICITED.

Boulton & Paul,

THE ORIGINAL MAKERS OF KENNEL AND POULTRY APPLIANCES,

NORWICH.

DUCKS' HOUSES.

DUCKS' HOUSE,
With Wire Netting enclosure, in water.

No 66.

REGISTERED COPYRIGHT.
Size—6 ft. by 2 ft. 6 in., 3 ft. high, with partition.
Cash Price, Carriage Paid,
£2 10 0

No. 66a.

Size—4 ft. by 2 ft. 6 in., with floor, ladder, and posts for standing in water.
Cash Price, Carriage Paid,
£2 10 0

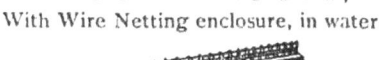

REGISTERED COPYRIGHT
Cash Price - £3 10 0
As Erected for Lady Lonsdale.

New Improved Ducks' House.

Cash Price, size, 3ft. 6in. by 2ft., 21/-.

SEND FOR
Illustrated Catalogue
FREE ON APPLICATION.

All orders amounting to 40/- nett, carriage paid, to the principal Railway Stations in England and Wales.

No. 50. Galvanized Duck Pond.

Cash Price, 10/6 each.

ADVERTISEMENTS.

The success of this Journal, which has so wide a circulation, enables us to offer Advertisers a medium for placing their goods before agriculturists and rural residents second to none. Unanimous testimony to this effect from those who have used its columns will be forwarded on application to intending Advertisers.

Farm Field & Fireside

Farm, Dairy, Live Stock, Stable, Poultry, Garden, Home.
A JOURNAL FOR EVERYBODY.

POULTRY	This department is conducted by W. Cook and has attained a widespread celebrity by the simple and practical manner in which Poultry-keeping, from a profitable point of view, is treated.
THE GARDEN	This important subject has been entrusted to a thoroughly-competent practical man, who has also attained marked success as a writer upon horticultural subjects. Those alike who utilise the Garden for pleasure or for profit will find the columns of "FARM, FIELD, AND FIRESIDE," a *vade mecum* for all their requirements.
THE HOME	The Fireside department is conducted by one of the most able lady writers of established reputation, and who is a recognised authority on all matters pertaining to the house. Articles appear weekly on "Our Recipe Book," "Novelties," "Passing Fashions," "Domestic Medicine," "Household Hints," &c., illustrated by well-known artists and engraved in the highest style of art.

QUERIES AND ANSWERS.

Especial attention is called to this feature of the paper, as the columns of every department of "FARM, FIELD, AND FIRESIDE" are freely open to all, and offer a means of exchanging opinions and obtaining information such as can be met with in no other way.

N.B.—"FARM, FIELD, AND FIRESIDE" offers a greater number of pages of well-printed useful information in a handy, compact form, illustrated, stitched and cut, for the sum of

☞ ONE PENNY.

Specimen Copies can be obtained from Newsagents, Booksellers, and Bookstalls, or direct from the Publishing Office,

1, ESSEX STREET, STRAND, LONDON, W.C.

ADVERTISEMENTS.

JEYES' FLUID,
(Non-Poisonous),

The Best and Cheapest Disinfectant, and stronger than Carbolic Acid. Invaluable for the House, the Stable, the Kennel, and the Poultry House.

SANITARY POWDER.
DISINFECTANT SAWDUST.

80 Prize Medals 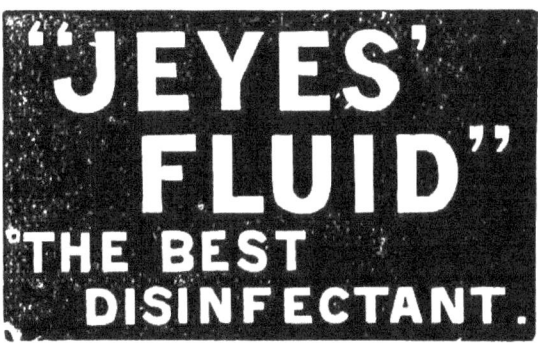 and other Awards

JEYES' FLUID destroys Fleas and Vermin of all kind.
Cures Roup, Gapes, and Comb Diseases.

Testimonial from W. Cook, late of Tower House, Orpington, Kent:—

"I find it excellent for disinfecting the poultry-houses. When there is a contagious disease raging it is most useful for sprinkling about, and in cases of Comb Disease, for outward application, I have found it surpass everything."

DOG SOAP. POULTRY SOAP.
In Tablets and 1 lb. Bars.

Testimonials, Price Lists, and all particulars on application to—

JEYES' SANITARY COMPOUNDS COMPANY, Ltd.,
64, CANNON STREET, LONDON, E.C.

DUCK MEAL.

I HAVE brought out a Meal especially for Ducks, but have not offered it to the public till now. I have used it myself for many years, and have every reason to believe that it is the best meal that has been manufactured for Ducks. I do not know anyone who has got his Ducks up to the weight at the age of mine. Perhaps there may be some, but I have not heard of any. I am sure those who feel inclined to give this Duck Meal a trial will be more than satisfied with the result. It mixes without sticking, there is no waste whatever, and the Ducks are very fond of it. It does either for rearing the young Ducklings or for stock Ducks, but I like to use biscuit meal the first fortnight the young ones are hatched.

The Prices for the Duck Meal are as follows:

14s. per cwt. ; 7s. 6d. per ½-cwt. ; 4s. ¼-cwt.

CARRIAGE PAID TO ANY COUNTY IN ENGLAND.

I may say it is made up of the most nutritious materials, including a great deal of meat, bonemeal, and oatmeal. I do not wish to imply that other Duck Meals are not good.

W. COOK & SON,

Queen's Head Yard, 105, Borough, London,

AND

ORPINGTON HOUSE, ST. MARY CRAY, KENT.

SPRATT'S PATENT CHICKEN MEAL,

See you get Our Meal in
SEALED BAGS,
Per cwt. 20s.; per half cwt. 10s. 6d.;
per quarter cwt. 5s. 6d.;
Per 14 lbs. 2s. 9d.; per 7 lbs. 1s. 6d.;
Or 3d. and 6d. SAMPLE PACKETS.

Samples of Foods
AND
Pamphlet on Poultry Rearing
Post Free for One Stamp from
Spratt's Patent Limited,
BERMONDSEY, LONDON, S.E.

Spratts Patent Limited
Bermondsey

Feb 27th/1894

Dr Sirs,

If my opinion of your Poultry Meal & Griessel can be of any service to you, I can confidently say that it has worked marvels in my Flock. Since I introduced it, I have been able to score a grand total of wins in the Show Pen.

I have tried many Expensive Foods, but I have no hesitation whatever in giving preference to yours. I reared over 200 chickens last year, and they thrived amazingly on your Meal. What I sold, were in grand condition and plumage, and realized remunerative prices & a quick sale.

For exhibition Ducks, it is unequalled, & so long as I remain a Waterfowl Fancier, so long will I use "Spratts Patent Food" for them.

Yours Resp.y
J. E. Evans
Sec to the Hanwich Dog & Poultry Soc.y